Boise River

Gold Country

BOISE RIVER

GOLD COUNTRY

Evan E. Filby

With input from Skip Myers, Boise Basin Merchant

Sourdough Publishing
Idaho Falls, Idaho

Cover: Prospectors search for "color" – gold dust or nuggets – along a mountain stream. If they find enough, they will set up a rocker or riffle box to begin production.

ISBN 978-0615624198

Boise River Gold Country is produced, upon order placement, by CreateSpace, an amazon.com company. Orders may be placed with the *Gold Country* eStore at www.createspace.com/3826870 or through www.amazon.com. We will consider bulk orders at special discounts for educational purposes, fund-raising, or corporate gifts. Museums, historical associations, and schools are particularly encouraged to apply. Contact the author through the imprint addresses provided below.

Sourdough Publishing
2184 Channing Way, Suite 437
Idaho Falls, ID 83404
e-mail: Sourdough@Earthlink.net

ACKNOWLEDGEMENTS

The idea for this book originated with Skip Myers, about three years ago. Born and raised in the Boise Basin, Skip left and then returned to start Donna's Place, now a café and grocery store. He is very concerned that they are losing their storied history as time, and old-timers, pass along. Of course, 2012 is the 150th anniversary of when prospectors founded many of the Basin towns, including Idaho City. Clearly, that would be a milestone to celebrate. However, existing books about Basin history were all out of print. Thus, he would have nothing to offer interested visitors and store customers.

After we made contact – through our respective sites on the Internet – the notion to put together a new book bloomed. Skip arranged a meeting with the Idaho City Historical Foundation (ICHF), and we proceeded from there. While I wrote the text and gathered photos, he informed me of some key activities and people that should be included. Thank you Skip!

We also need to acknowledge the work of all those who are trying, usually with minimal resources, to preserve our history. That obviously includes the ICHF as well as the Idaho State Historical Society. Other small county or regional historical organizations also deserve our support. Finally, we should include the U. S. National Archives, the Library of Congress, and the Smithsonian Institution.

On a personal note, I want to thank my wife, Caroline, who puts up with a hubby who sometimes ignores his surroundings when he's working on a tricky paragraph. I have little doubt that Skip appreciates the fact that his wife, Donna, puts up with his push to get this book published.

PHOTOGRAPHS AND OTHER ILLUSTRATIONS

I selected most of the pictures in this book to flow with the written history. Only for Chapter Ten – the recent and current situation – do the photographs link with the overall theme rather than specific content. I provide longer captions for those images.

We know dates for a fair number of photos, drawings, and so on, but that information is a bit hazy for some. Thus, *with confidence*, we can only say that most images are generally from the time period they are meant to illustrate. For natural features where records show not much has changed, I did substitute a few more-recent photographs.

In some cases, I had to use images from other Western venues to illustrate important Idaho activities. Fortunately, we know that the equipment, techniques, and clothing were much the same all over the West. In fact, it is not impossible to imagine that some individuals who appear in vintage photographs taken elsewhere may well have performed similar activities in Idaho. After all, many did.

Portrait photographers mostly recorded "old and honored" pioneers. Thus, pictures of gray-bearded older men often appear with stories that involve vigorous, physical contributions to Idaho history. Please make allowances.

Finally, it must be admitted that some vintage images have not survived "the ravages of time" as well as we might have wished. I have thus carefully applied Adobe® Photoshop® software to remove or minimize obvious spots, faded areas, and other blemishes. I realize that some may insist that the images be presented as is, "warts and all." However, I feel the reader should be allowed to benefit from modern technology that can capture potentially-important details and nuances. Please be assured, in view of certain notorious instances, that no individuals have been expunged from (or added to) any image used in this work.

Idaho City Stage

TABLE OF CONTENTS

SETTING THE SCENE

Historically, Idaho has been among the richest sources of precious metals any-where. Today, the state is still one of the largest silver producers in the world.

Boise River Gold Country
Heavy line, Boise River – South Fork through Featherville, Middle Fork to Atlanta, and North Fork above that. Dotted lines – Mores Creek through Idaho City, Grimes Creek through Centerville and Pioneerville.

But **gold** made Idaho a Territory. The richest source of the nuggets and dust that drove that event lay in what is called the "Boise Basin." The Basin is located in the mountains to the northeast of Boise, the state capital. The initial finds were along Grimes and Mores creeks. Pioneer City (later Pioneerville) and Centerville sprang into being along Grimes Creek. Idaho City formed along Mores Creek. Idaho City is the only county seat Boise County has ever had.

Compared to the rugged mountains around them, the Basin is more of a high, hilly plain. Not including the dividing ridge between the creeks, the region averages roughly 4,200 feet in elevation. That's about fifteen hundred feet higher than Boise. However, from these Basin towns, high ridges and tall peaks dominate the skyline in every direction. Three peaks stand over a mile higher than the Basin. Five more range from around a half to about three-quarters of a mile above the plain. Only the lure of gold drew white men into this forbidding landscape.

Confluence of the North and Middle Fork of the Boise river

Soon the Basin gold fields grew crowded or became depleted. Newcomers had to look outside Idaho's "Mother Lode." They found that other parts of the Boise River watershed also had gold-bearing placers and lodes. Thus, this book is not about just the Boise Basin. It also covers the entire mountain region drained by the tributaries of the Boise River, hence the title.

Programs and attractions based on historical materials, even living history sites, usually present sanitized versions of life during those times. Even presentations that keep some level of authenticity tend to feed us the "colorful" and "exciting" aspects. And one can hardly fault them. Who wants a steady dose of stories about how tough the pioneers had it?

Gold Miners Excavating Ore by Hand

So let's make that point right now. Life in the "Good Old Days" was hard, dirty, and dangerous. Most people faced brutally demanding physical labor. Accidents were common, and physicians were rare to nonexistent. Disease was rampant, and sanitation and medical practices were primitive, at best. You cannot fully appreciate what the pioneers did without considering those realities. I'll try not to dwell on those parts from here on. However, you should pause now and then to remember the hard facts of pioneer life.

The history that follows is arranged more or less in chronological order. Sometimes, when records are complete enough, events can be told as though you are looking over the shoulders of the participants. "You are there" scenes that are created as a composite of varied source materials are identified as such.

What follows is based on pioneer diaries and memoirs, early histories, period newspaper articles, and other factual materials. A bibliography of most-used references is included in the back.

CHAPTER ONE

BEFORE THE GOLDEN AGE

The column of men trailing out of the brush would have startled the Indians camped along what some called the *Soho'ogwaite,* Cottonwood River. They had heard of the *daiboo'nee,* pale-skinned strangers, from their cousins to the east and trading partners to the west. But no one along the river had ever seen one.

The travelers looked tired and bedraggled from the rain. Still, they displayed wealth beyond the Shoshone's wildest dreams. Most carried guns and metal knives. Their packs were loaded with goods the Indians had only heard tales of. They seemed friendly enough. Perhaps they would trade with the locals for some of their fabulous wares.

The natives along the river also impressed the EuroAmerican visitors. In his *Overland Diary* for November 21, 1811, leader Wilson Price Hunt wrote, "Some Indians who had pitched camp there had many horses and were far better clothed than those whom we had seen recently."

Humans have hunted what we call the Boise River watershed for many thousands of years. They included bands of Bannock and Shoshone Indians, with less frequent visits by the Nez Percés. Among the most notable of these groups were members of the Shoshone sub-tribe known variously as "Mountain Shoshone," "Sheepeater" Indians, or the *Tukadeka.* The latter name is commonly translated from Shoshone as "mountain sheep eaters." The name, as used by other Shoshones, honored the tribemen's skills as big game hunters.

Wilson Price Hunt

Never large in numbers, the *Tukadeka* ranged from the Hells Canyon area in western Idaho all the way east to the Wind River region in Wyoming. By and large, they stayed in the high country, including the highest ridges where the mountain sheep lived.

When EuroAmericans first met the *Tukadeka*, they marveled at the quality of their dressed furs and leather. Yet the *Tukadeka's* powerful bows impressed outsiders the most. Trapper Osborne Russell met a band and wrote, "They were well armed with bows and arrows pointed with obsidian. The bows were beautifully wrought from sheep, buffaloe and elk horns secured with deer and elk sinews, and ornamented with porcupine quills and generally about 3 feet long."

Curious Mountain Sheep

Witnesses reported that a *Tukadeka* bow could drive an arrow through the body of a buffalo. Some have claimed that other Indians might trade as much for such a bow as they would for the white man's rifle. That contention is lent credence by the fact that Indians could make and re-use arrows, while they had to keep paying for powder and shot.

A handful of today's best craftsmen have reproduced the *Tukadeka* horn bow. They created weapons that were as powerful and accurate as virtually any modern bow. Our high-tech materials stand up better to wet weather conditions than sinew and the old water-based glues, but that's about their only obvious advantage.

The *Tukadeka*, and other Shoshone and Bannock tribesmen, had the Boise to themselves until early in the Nineteenth Century. By then, bands in eastern Idaho had obtained horses and become skilled horseback hunters. The *Tukadeka* themselves, ranging the wild high country, found it impractical to keep horses. They used dogs as pack animals instead.

Indians along the Boise River would have heard stories about white explorers and traders in the Northwest. Most directly, they had ties of kinship with the Lemhi Shoshone who met Meriwether Lewis, William Clark, and the Corps of Discovery in 1805. They also traded with tribes in Oregon, who in turn dealt with American, British, and Russian trading vessels along the coast.

The party of fur trade explorers led by Wilson Price Hunt had just stumbled off the high ground to the southeast of the lower Boise Valley. Hunt was a partner in the Pacific Fur Company. The PFC was a subsidiary of the American Fur Company, owned by John Jacob Astor, a giant of the fur trade and one of the wealthiest men in America.

The "Overland Astorians," as they are generally known, had been sent to assess fur hunting prospects between the Rocky Mountains and the Pacific Coast. They were headed for Astoria, the base a sea-borne contingent had built near the mouth of the Columbia River.

The Astorians tried to navigate dugout canoes down the Snake River from eastern Idaho. Unfortunately, they suffered tragedy at "Caldron Linn," a stretch of deadly white-water near today's Murtaugh. They lost a man and many supplies and, unable to obtain horses, were walking to the West Coast.

J. J. Astor

After communicating with the locals, Hunt wrote, "They told us that farther upstream beaver were plentiful, though in the vicinity of our camp there were very few."

Caldron Linn

Astoria, ca 1813

The Overland Astorians finally arrived at Astoria in February 1812. The following year, John Reed, who had been with Hunt on his trek, returned to Idaho and established a post near where the Boise River flows into the Snake. From there, he dispatched parties to trap beaver along the Boise as well as other nearby streams.

Sadly, Indians – Bannocks, perhaps, or possibly *Tukadeka* – murdered Reed and his ten men. Pierre Dorion, Kentuckian Jacob Reznor, and another trapper were killed along the South Fork of the Boise in early January 1814. For many years after, trappers referred to the Boise as "Reed's River."

The only survivors of this venture were Dorion's Indian wife and their two young children. She struggled into Oregon through winter snows, then hunkered down in a rude shelter for a couple months. When the weather eased, they continued west until she contacted some other Astorians.

British encroachments during the War of 1812 ruined Astor's Western business. He dissolved the Pacific Fur Company in late 1814. Whites did not return to the upper Boise River region for another two years. Then, Scotsman Donald Mackenzie led a large expedition, referred to as the "Snake Brigade," into Idaho.

Mackenzie had also been a partner in the PFC, but now worked for the British-Canadian North West Company. Unfortunately, he did not keep a journal, so we know little about where he might have sent smaller parties in the Boise watershed. In 1819, warriors burned an outpost Mackenzie had built near the river mouth.

Donald Mackenzie

Alexander Ross

Not until 1824 do we have any record of whites in the upper Boise region. Three years earlier, the British government had forced a merger between the North West Company and the older Hudson's Bay Company. The Company had promoted Mackenzie to a post in Canada, so another Scotsman, Alexander Ross, now led the Snake Brigade.

Ross entered Idaho from Montana, reaching the Salmon River near the present city of Salmon. He decided to explore new territory and proceeded upstream. From there, they crossed over to the Wood River. The Brigade traversed the Camas Prairie and turned northwest into the mountains. They camped about eighteen miles north of today's Fairfield in early July.

They read the topography and turned west, which brought them to the South Fork of the Boise. In his *Journal* Ross wrote, "Never did man or beast pass through a country more forbidding or hazardous. The rugged and rocky paths had worn our horses' hoofs to the quick, and we not infrequently stood undecided and hopeless of success. However, after immense labour, toil, and hardship, we got to the river. Arriving on its

South Fork near Featherville

rocky banks, and looking, as it were, over a mighty precipice into the gulf below, we were struck with admiration at the roaring cataract forcing its way between the chasms and huge rocks over a bed it had been deepening for centuries."

Continuing downstream, with detours to avoid the worst stretches of river canyon, they next camped near where the future gold town of Featherville would appear. Brigade leaders normally sent scouting parties off from their main line of march. Thus, it's likely that trappers did explore the area along the Boise. However, their reports apparently did not encourage Ross, for he stayed near the South Fork until they could cross over onto the plains.

After his observations became common knowledge, the bulk of the Snake Brigade avoided the upper Boise watershed. For example, journal entries for the 1827-1828 campaign mention trapping along the Weiser and Payette rivers, but not the Boise. In fact, the Brigade deliberately swung further south to find an easier path across the mountains to the Camas Prairie, and then hurried east.

However, individual trappers went wherever they could find beaver. That included the Forks of the Boise, where – as in most Idaho watersheds – trappers practically wiped out beaver stocks. After that, few whites penetrated the region. Later emigrants on the Oregon Trail were interested only in getting across Idaho as fast as they could.

Emigrants on the Oregon Trail

CHAPTER TWO

GOLD RUSH CREATES IDAHO

August 1862: The column of eleven white men steered their horses through willow thickets and other shrubbery, sometimes splashing through the clear, fast-flowing stream that ran up the gorge. They remained alert against an ambush; several Indian bands lurked nearby. Earlier, a friendly tribesman had warned the leaders – Moses Splawn and George Grimes – that some parties contained "warriors of the worst type."

Finally, they exited the canyon into a meadow that was part of the Boise Basin. The flats widened to a couple hundred yards along the stream. After two miles or so, the watercourse jogged right around a higher shoulder. Perhaps another mile upstream, they decided to make camp.

Moses Spalwn

Around the campsite, hilly terrain rose slowly to low ridges a mile or two away. Only to the east did they see a loftier ridge – a thousand feet higher – about two miles away. Thick stands of fir, spruce, and Ponderosa pine, with scattered patches of lodgepole pine and aspen, painted the hills and ridges with green.

As they settled in, prospector David Fogus scooped up a shovelful of dirt and gravel, and swirled it in a gold pan. Within

Stretch of Grimes Creek

minutes, he had "color" – bits of gold dust and nuggets peeking from the last dregs of sand and small gravel. More followed. Splawn said, "I then felt we had found the basin of my dreams, so accurately described by my Indian friend."

Placer gold – that which can be washed out of sand and gravel – is scoured from ore beds by rushing high-country streams. Over millennia, fragments collect in "alluvial" deposits where the water spreads out onto flatter ground and slows down. Splawn had chased placer gold for nearly a decade and was a credible jackleg geologist. He and the other experienced prospectors in the group suspected they would do even better closer to the foot of the high ridges.

They marched upstream to another promising campsite. Here, peaks and ridges rising six hundred to a thousand feet or more brood over the meadow where they stopped. The creek, which had been rising a gentle 40-50 feet per mile, doubled its rate of climb toward a narrow cleft to the north. Within a mile of their campsite, a half-dozen smaller creeks flowed into what would be called Grimes Creek. Most of the nearby gulches showed promising color.

After two days, they decided to move even closer to what they hoped would be the local Mother Lode. Splawn said they climbed "over Pilot Knob and camped on the creek at noon."

Moses rode up a gulch, then left his hitched horse to climb a nearby prominence. At the top, he scrambled up a fir tree to get a better view. After checking out the lay of the land, Splawn clambered back down the tree. Then he saw something he had missed before. Not too long ago, an Indian had been around the tree, had perhaps climbed it to scout the countryside just as Moses had done. Alarmed, he plunged his horse through the thick brush, shredding his cloths and tearing his skin in many places.

He was resting from his ordeal the next day when the sound of shooting woke him. Grimes had just grabbed a shotgun, so Moses hefted his rifle and ran up the hill with him towards the rattle of gunfire. As they topped a rise, there was a volley of shots, and Grimes stumbled and fell. Splawn dropped beside him and no further shots came their way. With his dying breath, George said, "Mose, don't let them scalp me."

"Thus perished a brave and honorable man at a time when he stood ready to reap his reward," Splawn said.

Unsure how many Indians they faced, the miners buried Grimes in a deep prospect hole and fled the mountains. Only a large force would be safe from attack.

This Splawn-Grimes party was the first known white foray into the upper Boise River watershed in over thirty years. After about 1832, even the Hudson's Bay Company (HBC) trappers mostly avoided the area, except possibly short stretches of the South Fork. Individual trappers probably visited other streams, but they left no records. By 1842, the fur trade was a fading sideline. The HBC had built Old Fort Boise at the mouth of the Boise River in 1834. It now mostly did business with emigrant wagon trains along the Oregon Trail.

Old Fort Boise

As traffic along the Trail grew, so did Indian unrest. They naturally objected to the toll pioneer oxen and horses took on the best forage. Moreover, white hunting parties, armed with rifles, ravaged the supply of game animals. Small clashes culminated in the Ward Massacre of 1854. Eighteen of twenty pioneers in the Ward train were killed along the Boise, about 24 miles east of the trading post. After floods damaged it in 1856, the HBC abandoned the Fort.

By 1860, Oregon had become a state. Washington Territory encompassed all of the present state, plus Idaho and parts of Montana and Wyoming. Then the *Oregonian* newspaper, in Portland, reported (November 3, 1860), "A prospecting expedition has returned from the Nez Perces country. They report a rich gold region and Indians well disposed."

A company led by Captain Elias D. Pierce had discovered paying amounts of gold along Orofino Creek, a tributary of the Clearwater River in Washington Territory. Confirmation of the early reports set off a major rush. Of course, many of the early finds were located on the Nez

John Whiteaker, First Governor of the State of Oregon

Percé Indian Reservation. Authorities solved that niggling detail a few years later by "negotiating" a new treaty with the tribe.

Ignoring the legalities, around twelve thousand prospectors poured into the region within a year. The raw tent city of Lewiston mushroomed as a supply depot for the mines. Soon, the Territorial legislature carved out a new county in the region. As usual, events quickly outran politics. With so many miners searching, many found no claims left open. They began to spread southward, with new finds in the Salmon River drainage: Warrens, Florence Basin, and more. Toward the end of 1861, the legislature created yet another division – Idaho County – south of the first.

E. D. Pierce

Moses Splawn had joined the early rush into the Clearwater gold fields. Born in Missouri, he had emigrated with the family to Oregon in 1852, when he was seventeen. Within a year, he began what became a lifelong passion for prospecting. That brought him to Idaho in May 1861. He then followed the rush to Florence and did very well there.

However, Splawn proved to be more interested in the search than in making a fortune. He spent the hard winter of 1861-1862 in Walla Walla. By then, an Indian friend had suggested the Boise Basin as a good place to look for gold. Before he could pursue that, rumors of another rich possibility sent him on a search south of the Snake River in Idaho. That diversion is beyond the scope of this work, but by late summer Splawn and George Grimes had joined forces to explore the Basin.

As we saw above, the tragic death of Grimes aborted that expedition. (Conspiracy theorists have raised the possibility that another member of the party murdered Grimes. After 150 years there is no way such a speculation can be resolved.)

Splawn and the others hurried back to Walla Walla to gather reinforcements. In company with forty to fifty other men, they returned to the Basin. On October 7, they founded Pioneer City (today's Pioneerville) near the second Grimes-Splawn campsite at the foot of the higher ranges. The men spread out along Grimes Creek and soon established Centerville, near where Fogus had panned out the first flash of color. One member of their party was J. Marion More, who would become famous in Idaho mining history,

A few days after More and Splawn, another force of over sixty men reached the Basin. They were led by Relf Bledsoe, a renowned Indian fighter, who had been one of the first to establish a store at Florence. Seeking new ground, they founded Placerville 3-4 miles northwest of Centerville across a

J. Marion More

dividing ridge. Other camps were also set up, including a town called West Bannack on Mores Creek. (The place sometimes appears with the spelling "Bannock.")

Before heavy snow began to fly, hundreds of eager miners had arrived in the Basin. However, few were successful because streams froze over and there wasn't enough water to work the placers.

Life in the Western Gold Fileds, Period Lithograph

Even so, hardy men braved the severe weather to broadcast the news from the Basin. On January 12, 1863, the Territorial legislature split off the lower part of Idaho County, calling the new section Boise County. They made the log cabin town on Mores Creek the county seat. To avoid confusion with Bannack City in Montana, officials also tried to change the name. Locals rejected that, but a year later they finally agreed to call their town Idaho City.

Before the spring, an estimated seven to eight thousand pioneers had flooded into the Basin. They brought with them the tools and knowledge to work large claims. For that, they needed the rocker or sluice box.

The gold pan works all right to search for a possible gold-bearing sand bar or embankment. The tough steel pan can even be used directly to scoop up a sample. Then it's a matter of sloshing water around the pan to clean out the dirt and fine sand. Soon, you have a pile of stones, gravel, coarse sand, and (hopefully) denser gold that has settled to the bottom.

The prospector needs patience, and a certain level of skill to finish. Gold nuggets are easy, but it can be difficult to wash and flip the useless sand out of the pan without losing any of the fine gold. A group of miners might go directly to a sluice box, but a man

Prospector Using Gold Rocker, or Cradle

by himself usually turned to the rocker, or cradle. All he needed were tools he normally carried – axe and knife, and perhaps a chisel – and a length of log.

The base would have a flat top surface, with cross-ridges or cleats to catch the gold dust, and a round bottom so it would rock back and forth. Split slabs formed the basket suspended over the base. It had holes cut by a knife or burned through with a hot metal rod. No nails required, wooden pegs would do to hold it all together. Dump in a

Prospectors Using a Sluice Box. Sometimes Called a "Long Tom."

shovelful of dirt, sand, and gravel, and wash with buckets of water. Gold-bearing dirt and fine sand washed through and were caught by the riffles, while stones and heavy gravel stayed in the basket. Experience showed that one man could process probably five times as much material in a day, with much less effort, that he could with a pan.

A small team of men might work a larger cradle, with one digging ore while a couple more did the washing and separation. However, if the fall of the stream allowed, a sluice or "riffle" box was even more efficient. In its simplest form, the sluice box is a long open channel with cross-ridges (riffles) spaced along the bottom. The sluice can, in fact, be quite long, but that is not required for a manually-fed unit. The device does, however, require a continuous flow of water. The ore is shoveled into the upper end so it is washed thoroughly, the gold being caught on the succession of riffles.

Encouraged by the reports, Wells Fargo & Company established an office in Placerville. However, a chancy water supply curtailed work on the placers around that town by the end of May 1863. Work would not resume until locals found the resources to built a better system of ditches and flumes.

Flumes Bring Water for Placer Mining

Agitation to create a new Territory had begun even before the Boise Basin discoveries. Political leaders, elected mostly from districts west of the Cascades, were agreeable. All those new voters in the mining districts threatened their control of the Territorial legislature. Politicos disagreed only about where to cut the boundary of a new territory. They finally placed the border just west of Lewiston, which retained the maximum area for future population growth, but excluded all those miners.

Created on March 4, 1863, Idaho Territory included the future states of Idaho and Montana, plus most of future Wyoming. It was, in fact, much larger than Texas. Far behind events, Congress made Lewiston the capital, although the population in the north had already dropped precipitously.

Many gold seekers came simply hoping to make their fortune; they had no interest in "colonizing" the wilderness. James Alonzo Pinney, who arrived in Idaho City about the time the Territory was created, became a prominent exception.

James Pinney

Born near Columbus, Ohio, at age fifteen James accompanied his father on the Gold Rush to California in 1850. After two or three years as a clerk, he joined a pack train hauling supplies into southern Oregon. Pinney served in the military during the Rogue River Indian War of 1854-1856, then returned to the Midwest to visit relatives. After a year or so there, he returned to the packing business in Oregon.

In 1862, James followed the rush to Florence and other Salmon River camps, and on into the Boise Basin. He spent the hard winter in Oregon. In March 1863, Pinney packed in supplies to open a store in Idaho City. By September, when the first Territorial census was taken, the town had nearly seven thousand residents – more than Portland, Oregon at the time.

An enthusiastic theater-goer, Pinney also built the first Idaho City playhouse, which featured the latest traveling theater and comedy acts. A year after Pinney built his first store, President Lincoln appointed him Idaho City postmaster, a position he would hold for eight years. Twice burned out by the fires that swept through town – in 1865 and 1867 – James rebuilt and carried on.

By the summer of 1863, the Basin held an estimated eight to ten thousand people. The Idaho City area had benefitted from a better water supply than Placerville and other camps. Miners there had water through the summer in many places. Moreover, one company had enough flow to operate a "hydraulic giant" on Elk Creek, north and a bit east of town. A "giant" shoots water from a jet nozzle and can wash an entire hillside down into a long riffle box. Although we decry their destructiveness today, they were considered the latest in mining technology for their day.

The Portland *Oregonian* reprinted (May 20, 1863) an item from Walla Walla that said, "At Placerville and Bannock City there is

FIRST BENEFIT

Of the Popular and Gifted Young Actress Miss

Charlotte Thompson!

Our Best Citizens and Old Theatrical Patrons.

"CAMILLE"

"CAMILLE," THIS NIGHT ONLY.

This Friday Evening, Oct'r 17th, '82,

CAMILLE!

CAMILLE,   MISS CHARLOTTE THOMPSON

The drama *Camille* was one of many top shows that appeared in Idaho City during this period. We do not know if the company that featured Thompson is the one that performed in Idaho.

quite a spirit of rivalry on the subject of the location of the Capital of Idaho Territory. The idea that Lewiston has any claim is ignored altogether."

The placers were so rich that, according historian Merle Wells, during the initial season, "Claims that paid less than $8 a day were not even worked … " This at a time when cowboys made a dollar a day, with bunk and grub, and a skilled carpenter in the East might make $2 a day.

In late summer, another permanent Idaho pioneer,

Typical Hydraulic Giant

Jonas W. Brown, arrived in Idaho City. Jonas was born in 1825, sixty to seventy miles northeast of Columbus, Ohio. He found work in Keokuk, Iowa when he was seventeen years old. Hoping for better prospects in a new country, he moved to California in 1853. There, he prospected, did carpentry, and served in public offices: Deputy Clerk and then Deputy Sheriff.

In 1862, Brown followed the gold rush to Florence, where he again filled several public offices – including County Auditor and Deputy Treasurer – simultaneously. He moved to Idaho City in the fall of 1863.

Attorney Jonas Brown

During his time in public office, Jonas had read law on his own and also studied in various attorney's offices. Thus, by the time he arrived in town, he had been admitted to the bar. The census showed that at that time Boise County held over half the Territorial population. Brown would practice law in the county seat for nearly twenty years.

He had plenty of company. Including Jonas, Idaho City was then home to twenty-five to thirty-five attorneys. That was roughly the same as the number of saloons in the town. The medical profession was also well represented, with fifteen to twenty doctors, two dentists, and three drugstores.

Besides several score other businesses, with the usual number of saloons and gambling venues, the town also contained a mattress factory and four breweries.

At the end of September 1863, Idaho City saw the publication of the first newspaper in southern Idaho, the *Boise News*. The following year, the founders sold the business and the new owners renamed it the *Idaho World.*

By then, the new town of Boise City had grown on the river near the Army's Fort Boise, built in July 1863. Its favorable location in a fertile valley soon gave the town an air of permanence compared to the Basin. In the Basin, camps and towns bloomed or withered with every new gold strike.

One such discovery occurred on the Feather River, a tributary of the South Fork.

Press room. *Idaho world*

First located in early spring, the fields held over a thousand miners by the end of May. Many set up camp at a spot about thirty miles southeast of Idaho City. The September census recorded 560 inhabitants in what census officials called the South Boise district. The following year, the town of South Boise, later renamed Rocky Bar, was formally established there.

The new finds outside the Boise Basin led prospectors further and further afield. One party under John Stanley started out from the camps in the lower Salmon River mountains. They struggled through the incredibly rugged terrain into what we call the Stanley Basin. Disappointed at what they found, many from the group turned back.

Stanley and the others crossed over the local divide onto the Middle Fork of the Boise. There, they found promising amounts of placer gold. In need of supplies, they headed for Idaho City, which was about thirty-five direct miles to the west. Arriving in early August, they tried to keep their find a secret but only partially succeeded.

Faro was Popular with Miners and Businessmen

In no time at all, hundreds of eager prospectors poured into the trackless mountains. Most returned within a week or so, after a fruitless scramble through the forbidding terrain. Their angry reports helped keep the Middle Fork discoveries hidden for awhile.

Through accident or (perhaps) design, prospects in the Stanley Basin sounded more promising than those along the Middle Fork. Thus, when those in the know gathered supplies and headed east from Idaho City, many hopeful miners hurried on over the divide. Meanwhile,

Middle Fork Canyon

Stanley's insiders established the camp they called "Atlanta" along the Middle Fork. They mapped the best placer fields and lode sources before organizing a mining district in July 1864.

Only then did the outsiders catch on. Even then, the area was so isolated, and the country so rugged, that individual miners had little hope of striking it rich. Moreover, just living there was expensive – virtually all supplies had to be brought in by pack train.

It soon became clear that the narrow canyons and small sand-gravel bars offered little scope for placer mining anyway. Most of the wealth was buried in lodes of gold-bearing quartz. That required greater capital investment, including the cost of a mill to crush the ore so the gold could be extracted.

Twelve miles to the southwest over towering ridges, Rocky Bar endured similar restrictions. However, that camp was easier to reach, and saw much more activity after the spring of 1864. Many small companies built the "poor man's"

Pack Train, Atlanta

Water-Powered Arrastra

mill, known as an "arrasta." Used around the Mediterranean Sea since at least early Roman times, Spanish settlers first brought them to the New World in the 1500s.

The arrasta's core is a pit lined with flat rocks. Using hammers, miners break the ore into fairly small chunks and toss the pieces into the hole. Then one or two huge stones attached to a vertical axis are dragged over the ore to grind it. The miners themselves, draft animals, or a water wheel provide the power to turn the axis.

Processors then gathered the crushed ore, or added mercury directly to the pit, to extract the gold as an amalgam. They next heated the amalgam to evaporate the mercury and leave the gold. They also had to recapture the costly mercury. The process was extremely inefficient. Moreover, prolonged exposure to mercury vapor can cause severe neurological damage to humans and animals.

Productive recovery from Rocky Bar ores would have to wait for a wagon road into the area from the outside. Entrepreneurs completed that before the year was out.

Meanwhile, on the other side of the Basin, operators hauled a ten-stamp mill to a mine located a mile or two west of Placerville. It turned out $5 thousand worth of gold in its first week.

Ten-Stamp Mill. The core hammer set stands 18-20 feet high – about two stories

A stamp mill uses a set of weighted hammers attached to a horizontal cam. The camshaft was rotated by a belt drive connected to water power wheel. As the shaft turns, swivel arms lift the hammers so they can fall onto a bed of wet ore. (The material is kept wet to reduce dust.) Unlike the arrastra, a stamp mill pulverizes the ore into a fine slurry. In the early days, the slurry was also treated using the mercury amalgam technique.

Both lode and placer mining were booming in the Basin. Placer miners had learned to not waste a moment. When they had plenty of water flow, the mines

Miners Work a Town's Foundations

ran around the clock. The *Boise News* reported (April 30, 1864), "We counted more than thirty mining fires on Tuesday evening from a single standpoint in front of our office door."

Sometimes, the fires were even closer. Miners and townspeople fought over just where the underlying placer claims ran. Prospectors occasionally dug right up to foundations, and more. Several structures collapsed when miners undercut their walls.

Seeing the riches, some owners actually took up their floor boards and dug down to bedrock through the subterranean gravel. They took buckets of sand and gravel to the nearest available water flow, where some recovered as much as $16 per pan.

Idaho City also benefitted as a supply point and staging area for numerous other camps in the ranges to the east and south. That even included a difficult route to Rocky Bar. Traffic from Idaho City to Rocky Bar followed the so-called "Goodrich Trail."

Suitable only for pack trains, the trail first ran up about ten miles of easy grade along Mores Creek. Then it turned east across four ridges over the course of about forty miles. Two of these spines had climbs of around two thousand feet, while a third rose almost 2,500 feet.

At Rocky Bar, companies kept the arrastras grinding, or used small mills that could be broken down for a pack train. The "Elmore" (sometimes "Ada Elmore") produced some of the richest ore in the area. One of its investors was the man who gave his name to Nevada's famous Comstock Lode, Henry T. P. Comstock. He had sold off his

Henry Comstock

Freight Wagons on Mountain Road

interest in the Comstock before its value mushroomed. Hoping to boost his Idaho claim, he told anyone who would listen that the Elmore would rival the Comstock. The mine would do well in the long run, but Henry again sold out and moved on (to Montana).

In October 1864, the first major shipment of heavy equipment rolled into Rocky Bar. The freighters used the new "South Boise Wagon Road," completed by crews led by Julius Newberg. A correspondent told the *Boise News* that when the wagons arrived, "Long and loud huzzahs rent the air and made the welkin ring. All business was for the time suspended and everybody seemed loud in their praises of the energetic and thorough-going Newberg."

The road connected with Goodale's Cut-off – a branch of the Oregon Trail – at Little Camas Prairie. The Prairie is 40-45 direct miles southeast of Boise. The road had to first drop 700-800 feet to the South Fork, then run upstream to Featherville. From there, it climbed almost a thousand feet to cross a ridge before descending toward Rocky Bar.

One of Newberg's partners in the toll road company was Robert A. Sidebotham. Born in 1834 on the Ohio River in Pennsylvania (that

Attorney Robert Sidebotham

20

is, west of Pittsburgh), Sidebotham graduated from Oberlin College with a law degree. He moved to the West around 1856. He prospected in Colorado and California, and also taught school in Utah.

Sidebotham followed the rush to Elk City in 1862, when that town was still in Washington Territory. After spending the winter in Oregon, he helped explore the Boise River gold country. He teamed up with Newberg and a third man to obtain a toll road franchise, which the legislature granted in January 1864. Sidebotham then opened a law office in Rocky Bar. Over the next decade, he practiced law, invested in mining properties, and "held every office that was in the gift of the people" of the county.

Even then, prospectors could see that the Boise gold country did not have enough claims for the thousands who were still pouring into the area. Fortunately, gold had also been found in the Owyhee Mountains of southwest Idaho. That, and subsequent silver finds, drew many prospectors away from the Basin.

With all this going on, a major problem remained. The new Territory of Idaho was proving to be almost ungovernable. Most obviously, winter snows cut off the capital at Lewiston from the rest of the Territory. Thus, on May 26, 1864, Congress created Montana Territory. Idaho Territory retained a part of western (future) Wyoming, and Lewiston was still the capital. That, however, was about to change.

Boise City had quickly become a major crossroads for the Oregon Trail and on the main route between the Basin and Owyhee mining towns. Moreover, the village had an air of permanence lacking even in the largest mining towns. Many of those towns

Idaho's First Capital Building, in Lewiston

Boise City, 1864

were little more than sprawling camps. They might literally disappear overnight upon rumors of another gold strike somewhere else.

In July, Boise City acquired a newspaper, the *Idaho Tri-Weekly Statesman.* Two months later, the editor trumpeted (September 17, 1864) Boise's advantages: "It is near the center of population, accessible at all times, with the advantages of a good climate, surrounded by a rich agricultural country, with untold and yet undiscovered mineral resources, there cannot but be a bright future for Boise City."

The next Territorial legislature agreed. They moved the capital from Lewiston to Boise City, effective December 24, 1864. After that, as predicted by the *Statesman*, the city grew steadily along with the Idaho mining industry.

By the end of that year, prospectors had located most of the important fields and lodes in the Boise River gold country. Ironically, history would quickly leave little scope for the individual miner, often referred to as a "sourdough" for his camp bread. Associations would form to pursue large-scale placer mining. Also, companies with greater financial resources would build or expand ore mills as lode mining increased. Large scale gold mining in the region would continue, off and on, for almost ninety years.

CHAPTER THREE

COOPERATIVE MINING REPLACES THE SOURDOUGH

Sumner Pinkham

The Warm Springs Resort lay quiet on a hot Sunday afternoon, July 23, 1865. The witness felt no particular premonition of trouble. But everyone knew about the bad blood between ex-sheriff Sumner Pinkham and gambler Ferdinand Patterson. Both had been in the spa's bar earlier, before "taking the waters" in the thermal baths in the back. Now Pinkham stood on the veranda, waiting for a ride. The Resort's hackney should be here soon, to carry him back to Idaho City, a mile or so east.

The witness only sensed danger when Patterson appeared on the veranda. He walked toward Pinkham and rumbled, "Will you draw, you Abolitionist son-of-a-bitch?"

Leastwise, that's what the viewer thought he heard. Before he could fully comprehend what was happening, shots rang out. Hit twice, Pinkham slumped to the deck, dead. The ex-sheriff had fired once, a bullet that went: Who-knows-where? Patterson fled with a band of cronies, but soon surrendered to the law.

Witnesses later agreed that Ferd had indeed called Pinkham an "Abolitionist son-of-a-bitch." The first "charge" was true enough, of course. A native of Maine, Sumner was both a strong Unionist and an ardent anti-slavery man. Patterson, reportedly a native of Tennessee, bitterly hated the recent dissolution of the Southern Confederacy.

At the subsequent trial, witnesses told radically different stories about who challenged who, and who had drawn first. In any case, Southerners who had fled West during the Civil War or

Ferd Patterson

23

Freighters in Idaho City

at the beginnings of Reconstruction dominated the Boise Basin camps. As expected, Patterson was declared "Not Guilty" by reason of self-defense. When Ferd was shot to death in Walla Walla the following February, many assumed a vigilance committee had sparked the deed. The shooter later somehow walked away from jail and disappeared.

When this incident happened, Idaho City was still by far the largest town in Idaho Territory. (If fact, it was still the largest in the Pacific Northwest.) That preeminence arose mainly from a lucky accident of weather and geography. Their water supply, necessary for large-scale placer mining, had been generally larger and more reliable than in the Grimes Creek drainage. Idaho City thus became a major supply depot for the Basin. Construction of a more direct road to Boise City enhanced that status.

Freighter John Carpenter

One of the freighters who hauled goods into the Basin was John R. Carpenter. An early Idaho history said, "Mr. Carpenter had seen the streets of Idaho City so congested with teams that it was almost impossible to make one's way among them." It also noted that he had, "seen as many as four dead men in the streets of Idaho City at one time."

Born in March 1846, between Schenectady and Albany, New York, Carpenter came to Idaho City in the spring of 1863. Along the way, he drove one of the family's wagons, towed by oxen. His father became ill with "mountain fever," so John used their animals to haul logs and lumber to building sites in the town. With a stake from that, his father bought supplies in Oregon and brought them into the town.

After a couple years freighting, John's father homesteaded a ranch in the Boise Valley. John worked there for a time, but then returned to freighting and packing. He also drove a stagecoach. In fact, Carpenter traveled in and out of the Basin for another fifteen to twenty years.

Its status as the seat of Boise County, and as a supply depot, gave Idaho City a sustainable economic base not shared by other Boise Basin camps. Also, in 1863, Roman Catholic Fathers Toussaint Mesplie and A. Z. Poulin built the first church in the Basin. Besides St. Joseph's in Idaho City, they established smaller churches in Placerville, Centerville, and Pioneerville. That Christmas, Father Mesplie held successive Catholic Masses at the three western churches. Meanwhile, Father Poulin held a regular and later a midnight Mass at St. Joseph's.

Father Mesplie

St. Joseph's Catholic Church, Idaho City

In 1865, a terrible fire swept through Idaho City, but desperate efforts saved the Catholic church. Unfortunately, it did not escape another fire two years later. Fire fighters saved some church materials, but the structure itself was a total loss. Undaunted, the good Fathers rebuilt at the same spot. That St. Joseph's Roman Catholic Church is still in use today.

With hordes of prospectors ready to claim every stretch of ground along even the smaller streams, the "easy" gold played out quickly. Soon, miners found they had to work their way further from the creeks and go after gold buried in the hillsides. That could not be done economically by men laboriously digging with picks and shovels. Experienced miners followed the lead of the company that had assembled a hydraulic giant on Elk Creek in 1863. Soon, operators began attacking entire hillsides with batteries of water jets.

Giants Working Hillside Near Idaho City

As noted before, placer gold does not just magically appear. It has to come from lodes – generally masses or threads of metal intermixed with quartz. Lode mining seemed to have gotten off to a relatively slow start near Idaho City and up Mores Creek. Still, by the late 1860s, prospectors had found a few rich ore bodies and were extracting valuable amounts of gold-laced quartz. For a time, miners had to reduce the ore in a hand mill, or send it elsewhere for processing.

Gold fields around Centerville and Pioneerville followed much the same pattern as those around Idaho City. However, that area had a generally shorter water season, so it took longer to deplete the rich, easier gravels. An item from the *Idaho Statesman* showed how that was not an unmixed blessing. The *Statesman* said (April 30, 1868), "We learn by a private letter from Pioneer City that water is growing scarce there, and many men are consequently out of employment. It is also the same at Granite Creek. Hands can be readily obtained for four dollars per day, and not sufficient work for half that is offered."

Early Pioneerville

Even so, toward the end of the 1860s, the "bloom" had gone off the Grimes Creek placer fields. The editor for the *Idaho World* at the time, James O'Meara, opined that "Pioneer has seen its best times and … is rapidly on the decline in trade and mining importance." O'Meara was hardly an unbiased observer, being generally a booster for Idaho City.

Of course, yields declined only in comparison to the fabulous amounts extracted before. The Basin still produced several million dollars worth of gold annually during this period. A better water year in 1869 made for a very good season. Operators had also begun to see some decent returns from lode mining. Processing ore from mines around Placerville and Quartzburg kept mills in that area fairly busy.

James Henry Hawley, later famous in Idaho law and politics, was among the early prospectors around Quartzburg. Born in Dubuque, Iowa, Hawley lost his mother at a very early age. He apparently did not get along well with his father. Thus, when his mother's brother moved to California, James went with that family. From there, he followed the initial gold rush to Northern Idaho in 1862. He was just fifteen years old. After trying his luck in Orofino and Florence, he found his way to the Boise Basin in May 1863.

For awhile, Hawley worked for wages at the Gold Hill Mine, near Quartzburg. He then purchased a placer holding and searched other fields to make his own claims. He did well enough to finance a return to San Francisco, where he attended City College. He also read some law on the side. He resumed mining in Idaho around 1868, while continuing his law studies. Boise County voters elected him to the Territorial House of Representatives in 1870. That began his long and successful political career.

Young Lawyer Hawley

The district's rich quartz ore encouraged investors to haul in more stamp mill capacity. Starting around 1869, mills would operate almost continuously for nearly forty years.

Far to the southeast, investments in mill equipment also helped the Rocky Bar lode mines do better. In fact, by 1866, that area had more mill capacity than any other region in the Boise River gold country. However, managers still tended to focus on the ores that offered quick returns. Thus, mining went in fits and starts as workers found and depleted the best ore pockets.

Ore Mills, West End of Rocky Bar

South Fork placer mining offered rather small returns in those years. As a result, Chinese miners played a much larger early role. (Typically, as placer "diggin's" became depleted, Anglos abandoned them or leased them to Chinese miners. These hard workers were willing to accept the lower returns.) The *Idaho Statesman* reported (May 9, 1868), "Placer mining is being vigorously carried on. Water is plenty. Several hundred Chinamen will soon be at work."

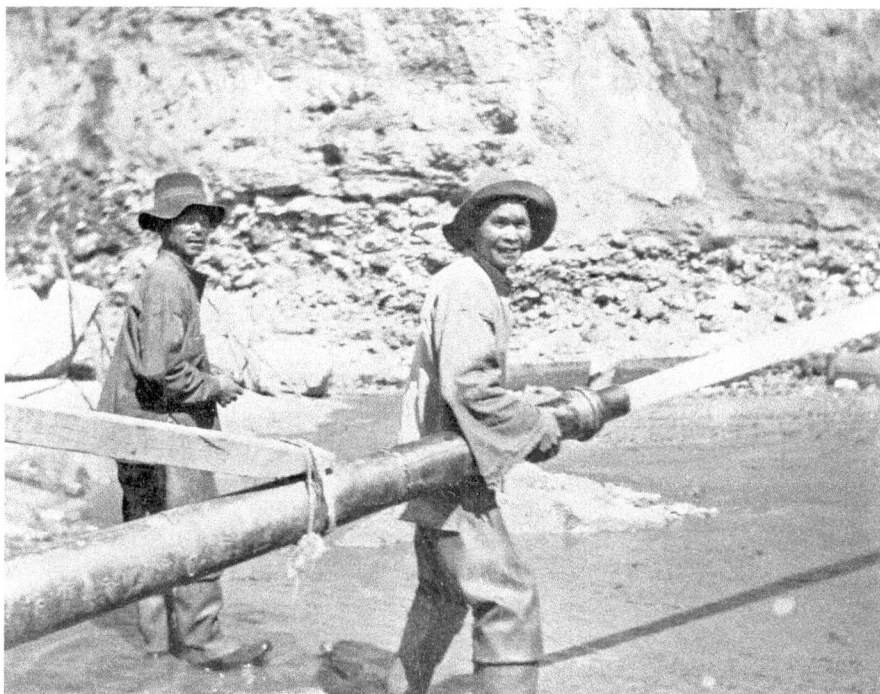
Chinese Operate a Giant Near Rocky Bar

As the only area town connected to the outside by a wagon road, Rocky Bar served as a supply center for the region. Thus, when the legislature created Alturas County in 1864, it selected Rocky Bar as the county seat. With growth tied to the mines, the town had its ups and downs. Some notable mining failures caused a hiccup in 1867. Short-sighted or dishonest management often caused such losses.

The town began to recover the following year. The May 9th *Statesman* article quoted above also said the lode mines were short-handed and that "There is a great scarcity of vegetables, and also a big demand for 'hen fruit.' ... " The following year proved even better. Another *Idaho Statesman* article (August 17, 1869) noted that most of the arrastras in the area were running: "All are busy crushing ore that will pay from $30 to $100 per ton ... "

Naturally, during lulls, mill owners found themselves with excess capacity. To keep their stamps busy, some of them invested in mines

Pack Train on Trail Near Atlanta

Chinese *Families* were Rare in the Basin

around Atlanta. Because of the area's isolation, exploitation of those and other Middle Fork lodes started more slowly. Arrastras were in use early, but not until 1867 did Atlanta have a stamp mill. (The owners had to break it down into pieces small enough for pack animals to carry.)

Atlanta quartz proved to be extremely rich, laced with much silver as well as gold. However, the processes then current in Idaho could not efficiently recover the silver. In fact, they basically threw it away. Even so, some of the richest Atlanta ores were profitable for a time. In 1870, owners of one of the best mines tried to sell the property based on the rich ore prospects. That attempt foundered after some early promise. The lack of all-season access seems to have been a major factor blocking the deal. In general, Atlanta mines would struggle to turn a profit for years to come.

With the switch to lode mines, and mechanized (hydraulic) placer mining, the population of the Boise River gold country declined sharply. The 1870 census recorded about 3,500 people in the Boise Basin. Nearly half (1,700) were Chinese. In fact, the total Territorial population decreased by about 7 percent (to around 17,800).

As the mid-1870s approached, renewed investor interest sparked a weak revival in the Atlanta mines. The ores simply seemed too fabulously rich to ignore. But the wastage in ore processing remained a major drag on profitability. At one point, operators even sent mule trains of ore to the railroad station at Kelton, Utah. From there, they were shipped to processing plants that could handle the complex materials. However, that was only a measure of their desperation. Even that was finally stopped.

Lode miners around Rocky Bar faced some of the same problems, except they were not as isolated from the outside. However, while their ores could be processed more easily, they were not as rich in precious metals.

Mules Pulling Freight Wagon

Miner T. L. Johnston.

Thomas L. Johnston had one of the better Rocky Bar lodes. Born about thirty miles southwest of Canton, Ohio, Johnston followed the lure of gold to California in 1853. He was just twenty years old. Thomas then prospected fields all around the Pacific Northwest. He came to northern Idaho after the gold discoveries there.

In October, 1862, he was among the men who marched into the Boise Basin under the leadership of Relf Beldsoe (mentioned in Chapter Two). After a few false starts, Johnston settled around Rocky Bar. There, he earned steady returns from various quartz claims. The *Owyhee Avalanche* reported (October 6, 1875) on a strike that made "principal owner" Johnston optimistic. It said, "The rock generally has a very productive complexion, displaying well defined streaks of free gold ... "

Early Rocky Bar

Like Johnston, operators sought higher-paying lodes. Many opened up new ledges and ran their shafts deeper and deeper. They also installed new equipment – better lifts, pumps, and more – trying to reduce costs. None of that worked particularly well, but at least a minor level of lode and placer gold production sustained the town.

Meanwhile, lode mines on the other side of the gold country in Placerville and Quartzburg were doing very well. A correspondent to the *Idaho Statesman* wrote (August 8, 1873), "While it is quite true that Boise Basin to-day is not the Boise Basin of 63-'4, it is nevertheless also true that at no time since the discovery of gold in the Basin has the future been any brighter than at the present."

Early Quartzburg

Still, placer mining in this region suffered from the vagaries of weather. A letter writer to the *Owyhee Avalanche* said (June 13, 1874), "The season here has been short, and very unprofitable so far as our placer mines are concerned. Notwithstanding the brilliant prospects in early spring, many of our best claims have done nothing, while a few have made only short runs."

But another *Avalanche* item said (June 13, 1874), "While our placers – rich as they are – have been unavailable this season, our quartz interests are quite flattering, giving great encouragement to business."

The 1873 letter writer to the *Statesman* would have agreed, noting many new lode discoveries. The correspondent went on, "Here is located the mining works of the Gold Hill Mining Co., Messrs. Mootry and Coughanour."

William A. Coughanour was born in southwestern Pennsylvania. He moved to Quartzburg in 1870, at the age of twenty, joining his Uncle David. He started out as a clerk, but moved quickly to a more responsible position. Coughanour would remain a stockholder in the Gold Hill for at least another quarter century.

Mining Manager William Coughanor

Miner Working at Tunnel Face

In addition to the Gold Hill, several new ledges had been opened up. Some existing mines had been drilled deeper or had branched into wider strata. As noted in the introduction, before power tools this required back-breaking, dangerous work. Each chunk of laboriously broken rock was hauled out of the tunnel, usually on someone's back. Then the miners had to brace the ceiling with timber frame. Even so, rock falls were all too common.

A few companies planned to add new milling capacity. Another *Owyhee Avalanche* article identified (November 2, 1875) the best lode mining sites and also said, "J. H. Hawley's diggings, in the immediate vicinity, are turning out well, and will realize handsomely for the owner."

Recall that James Henry Hawley got his Basin start at the Gold Hill. After his term in the Territorial House, voters sent him to the Council in 1874. (A Territorial Council is roughly equivalent to a state Senate.) At that time, he still lived in Quartzburg. He was married there on July 4th, 1875. The couple moved to Idaho City three years later.

As noted earlier, water supplies in the Idaho City area were better during this period than those along Grimes Creek and further west. Still, severe weather sometimes delayed the spring runoff, and therefore the start of the placer mining season. The *Idaho Statesman* quoted (April 8, 1871) the *Idaho World,* which observed that companies were just then getting started. That was roughly a month behind normal. Still, the item said, "Within two weeks all the principal Companies in the camp will be at work, and our city, at night, will be illuminated by the pitchwood fires of the miners, on the bar, hills and gulches for miles around."

To some, the future of Idaho City seemed secure. That same summer, a merchant began construction of a truly permanent structure for his general store. Clearly, the owner had experienced, or at least heard about, the fires that had devastated the town in the Sixties. The exterior walls of his new store were made from local brick. Then, at considerable cost, the builder imported huge iron doors, a half-ton each, to cover all the ground-level openings. Finally, workers filled the attic space with a three-foot layer of packed dirt.

1871 Fire-Proof Building, now Boise County Courthouse

This building would go through a succession of owners and uses before being purchased by Boise County. The historic structure is still being used today as the Boise County Courthouse.

Another sign of optimism was the decision by German emigrant Nicholaus Haug to expand his brewery business. Born in Württemberg, Haug had come to the United States in 1852, when he was just sixteen. He then made his way across the continent to Oregon, where he worked on a farm for some time. He did well enough to buy, or claim and prove, a property of his own. Then the lure of gold drew him to Rocky Bar.

A brief stint in the mines convinced him that was not for him. He moved to Idaho City

Brewer Nicholaus Haug

and went to work in a brewery. Finally, in 1868, he bought into the firm. He became sole owner within a few years and gained a reputation for "making a splendid quality of beer." In fact, the *Owyhee Avalanche* said (November 3, 1875), "Nikolaus Haug of the Miners' Brewery keeps the best beer in the Territory. His establishment is quite a favorite resort for the gentlemen of Idaho City and vicinity."

Oddly enough, James Pinney read the situation differently. He was a "veteran" of the town's earliest days, when thousands of prospectors thronged the streets. He also saw the transition to hydraulic giants, where a handful

Miners' Brewer & Bakery Token

of men could replace scores, forcing prospectors to look elsewhere. A population of less than a thousand in 1870 surely limited the chances for civilized amenities, like the theater. That year, he traveled to Boise City to open a book store, which thrived.

Three years later, Pinney resigned his position as Idaho City postmaster and moved to Boise City. Historian Hiram T. French would dub Pinney "the father of modern Boise" because of the improvements he pushed during five terms as Mayor. As a private citizen, he also insured that Boise would have the latest in genteel entertainment. He passed away there in February 1914.

Hydraulic Giant Cuts into Tree Line

IOOF Hall, Idaho City

Another event on the permanence side was the dedication of an Idaho City meeting hall for the Independent Order of Odd Fellows, Pioneer Lodge Number 1. The *Idaho World* wrote, "The Dedication … was a perfect success in all its features, and was without doubt the grandest affair of the kind ever witnessed in the Territory."

IOOF Hall, Interior

Pioneer Lodge Number 1 was the oldest IOOF section in Idaho, first organized in 1864. Idaho City no longer has its own lodge. Even so, the Hall dedicated in 1875 still hosts meetings of the Order on an intermittent basis.

Those who believed in a solid future generally based their hopes on the number of new quartz prospects being discovered in the hills around the town. Newspaper reports told of a substantial lode four or five miles to the northeast. Although minimal development had been completed, test samples from the mine contained "very rich ore."

Another firm was exploring the so-called Keystone ledge, a mile or so south of Idaho City. The *Owyhee Avalanche* reported (December 15, 1875) that, "One of the parties interested is now in San Francisco with samples of the ore to have it tested to show the moneyed kings there what they have … " If the samples were as rich as they thought, the company hoped "to sell for a handsome figure."

However, despite that optimism, Idaho City continued to lose population over the next several decades. An earlier issue (November 3, 1875) of the *Avalanche* had summarized the switch from placer to lode mining. The article said, "While [the placer mines] … have been gradually diminishing in productiveness for the past five years, quartz has been slowly coming to the front … "

The town declined because, for various reasons, the "slowly" part of that equation turned out to be much slower than expected. Perhaps only its position as county seat allowed Idaho City to hang on.

Arrows Point to *Some* of the Gold in this Quartz Sample.

CHAPTER FOUR

PLACER MINING FADES, LODE MINING GROWS

The stagecoach rattled out of Idaho City in the early afternoon of Wednesday, August 2, 1876. Traveling correspondent William Goulder ignored the dust, and the bumps, jerks, and clatter. He wanted to see the countryside. The first three miles or so rose steadily up a narrow valley flanked by alpine forest.

At the summit, Goulder saw a crew working on a half mile of wooden flume. Channel & Company needed water on the Grimes Creek side of the ridge, so they had dug ditches and built flumes to redirect flow from higher on the Mores Creek side. They clearly expected excellent returns to resort to such a costly strategy.

Stagecoach Headed Along a Draw

Reporter W. A. Goulder

At the summit, the coach turned north along the ridge. The top was narrow, but still wide enough for the road. The track ran along relatively level terrain for over a mile. Off further north, impressive mountains provided a magnificent backdrop. Goulder knew he'd need to restrain his enthusiasm. Idahoans had grown rather jaded about "splendid mountain scenery."

They rolled into Centerville about 4 o'clock. Plenty of time to stretch his legs and get a feel for the town. The column he'd written about Idaho City would arrive too late for the Thursday edition of the *Idaho Tri-Weekly Statesman,* but should appear Saturday.

Idaho City was no longer the "wild and woolly," wide-open camp of its early days. Goulder, in fact, described it as "quiet and orderly." He also said, "I arrived here too late to see much of the operations in placer mining ... "

As noted in the previous chapter, miners now had to go higher onto the hills and ridges, away from the stream bed. That made it even harder to get water onto their claims. Sometimes men had to haul loads of sand and gravel down to where they had set up riffle boxes along a stream. Owners still found gold, but the work was slow. For many of them, low summer stream flows ended their seasons.

Lode mining also proceeded slowly. Continued prospecting had uncovered and proven more and more ledges of gold-laden quartz. But, even more so than placer mining, owners needed money to develop their claims. The pool of men willing to work underground in hard-rock mines was relatively limited. And, for that time and place, they commanded premium wages. Timber supports had to be cut and dragged to the mine to shore up weak spots. Then the ore had to be hauled to a mill for processing.

Unfortunately, the sins of the past punished current owners. In those heady by-gone days, speculators and scam artists had sold too many empty holes, or no holes at all. Although there are few reports of "salting" a claim – burying nuggets for the suckers to find, for example – that was a fairly common scam in the West. There's no reason to believe it didn't happen in the Boise Basin. Sometimes swindlers would buy, or steal, ore from a rich claim and pass it off as their own.

Heavily Timbered Mine Tunnel

Thus, outside investors had grown wary, making development funds hard to come by. Owners had to rely on their usually-meager personal resources. Most worked right alongside the handful of men they could afford to hire. Goulder wrote, "Their operations must be necessarily restricted and their progress slow ... "

Group in Front of Centerville Saloon

He found Centerville much the same. Some placer areas had water enough for a long season. Much more of the ground did not. The owners of those gold fields had to spread their mining over a series of short seasons. Thus, Goulder wrote, they could "give employment to several hundred men for several years to come." (Of course, those men would need to seek other employment in the off-season).

When Goulder visited Centerville, no one was doing much with the quartz mines in the area. In fact, he described the town as being "but the wreck of its former self, having been swept by fires and only partially rebuilt. The business houses are just one of each kind."

Thus, he stayed at the only hotel and, no doubt, patronized the only saloon. Goulder does not mention the only livery stable in town, run by James Pulk and Cary Havird. Little is known about Mr. Pulk. Havird was born in Quincy, Illinois, along the Mississippi River. His family moved to Centerville in 1865, when Cary was ten years old.

Growing up among prospectors, he mined some himself as a teenager. However, he found driving a freight wagon or working on a ranch more congenial. In 1875-1876, he teamed up with Pulk to run the Centerville livery stable. Ten years later, voters elected Havird to the position of Boise County Sheriff.

Liveryman Havird

Further up Grimes Creek, Pioneerville was more active and prosperous than Centerville at that time. A number of rich placer claims made much of the difference.

One long-time pioneer, Ben Willson, had extensive holdings back among the ridges as well as near Grimes Creek. Goulder observed Ben's claims during his trip through the region. He noted that at one of his claims Willson "was working two monster hydraulics." Of course, that required a large, expensive system of ditches and flumes to supply a strong water flow. Clearly, Willson's claims were paying well.

Willson's Hydraulic Apparatus

Although not as extensive as those around Idaho City, a few quartz claims helped sustain the Pioneerville economy. However, also as in Idaho City, the owners lacked capital to fully exploit their holdings. In fact, that problem would plague both placer and quartz miners in all three towns – Idaho City, Centerville, and Pioneerville – until at least the mid-1880s.

Conversely, the lack of capital proved less of a problem for the lode mines to the west, around Quartzburg and Placerville. Prospectors opened few new mines during this period, but the known lodes produced steadily.

In 1882, a reporter visited the Gold Hill Mine, where William Coughanour was still superintendent. The writer observed (*Owyhee Avalanche*, June 24, 1882) that "The ledge has been worked for about fifteen years and has paid handsomely … "

In an earlier column, the *Avalanche* correspondent wrote, "I took a seat by the side of Bob Curtis (driver of the U., I. & O. stage from Boise to Idaho City) and was soon on my journey to the rich placer camp[s] of Boise Basin."

The Utah, Idaho, and Oregon Stage Company ran the coach into Idaho city. The route continued on to Placerville and Quartzburg. Tennessean John Hailey was the principal owner of the UI&O line. He had long been a pioneer in the Basin and, indeed, large parts of Idaho.

Born in Smith County, 30-40 miles east of Nashville, Hailey emigrated to Oregon in 1853. He was then eighteen years old. John left work as a miner for distinguished duty in the Rogue River Indian War of 1855-1856, then went into ranching. With the discovery of gold in Idaho, Hailey sold his holdings in 1862 to finance a company shipping supplies into the Basin. As the roads improved, his firm switched from pack trains to wagons.

Delegate Hailey

In about two years, John began a stagecoach service that connected the various Basin camps to Umatilla, Oregon. One of the most efficient operators in the West, John's company prospered. In 1870, he sold it at a substantial profit and settled down in Boise City. There, he ran a thriving meat market, supplied from a ranch he owned outside the city. In 1872, voters elected Hailey to be Territorial Delegate to the U. S. Congress. (Delegates have no vote on the floor, but can serve on committees and vote on issues at that level.)

His term so impressed people in Idaho that both political parties urged him to take the position again. He declined, and returned to his business interests. Shortly thereafter, John had some financial reverses. Perhaps worst of all, he had to make good on several notes he had co-signed. To recoup his losses, Hailey went back into the stage line business. He organized the Utah, Idaho, and Oregon Stage Company in 1878.

John again did well. He eventually branched out to buy land and mining claims in the Wood River Valley. The present town of Hailey is named for him. In 1884, he served another term as Delegate to Congress. Two years later, he dissolved the stage line because of competition from Idaho's new railroads. For awhile before that, the UI&O also ran a stage into Rocky Bar and Atlanta.

Stagecoach Stop in Boise City

The first wagon road reached Rocky Bar in October 1864 (Chapter Two). But attempts to provide stagecoach service repeatedly failed. The track was simply too rough. Finally, in 1870, W. Cornelius "Con" Tatro decided the road had been improved enough. He initiated regular stage service.

Con Tatro was born in northeastern New York State in 1838. The family moved to Wisconsin in 1849 and then, after four years there, on into Minnesota. Con and his brother Joseph, who was three years younger, arrived in Idaho in 1863. They had apparently spent two or three years mining in Colorado before then.

We know that Joseph spent over a decade prospecting in the Boise Basin. Con probably did too, and he became familiar with the Atlanta-Rocky Bar region. He may have carried the mail for a time before his stagecoach venture. From the pen of William Goulder, again, we know that Tatro's stage connected with the UI&O at "Rattlesnake Station," not far from today's Mountain Home.

Freighter Joseph Tatro

The success of Con's line encouraged Joseph to also leave mining. He focused more on hauling freight. Joseph eventually had routes from Kelton, Utah into Rocky Bar and then on into Oregon.

In 1880, Con Tatro extended his stage line to Atlanta. He sold the business to John Hailey two years later and became the town marshal at Hailey. Joseph got out of the freight business in 1886. Like John Hailey, he surely did not want to compete with the railroad. He then went into stock raising in Cassia County.

When Hailey dissolved his company, one of his partners purchased the equipment needed to maintain service into Rocky Bar.

Early Atlanta

Like other Idaho gold areas, the Rocky Bar and Atlanta mines had been plagued by scams. These had made potential investors suspicious. As a result, mines here ran in fits and starts, when financing became available. A writer for the *Owyhee Avalanche* noted (December 8, 1877) the end of one lull. He said, "The mining interests here have assumed their usual appearance, and confidence in the future has returned to our people ... "

He went on to say that the old Vishnu Mine – located about a mile west of Rocky Bar – was again producing at a good rate. He wrote, "The mine is improving as the main drift advances on the lode."

By around 1876, prospectors near Atlanta had found surface signs of a major out-cropping. It stretched for about two miles. However, the *Owyhee Avalanche* reported (December 16, 1876), "No developments have been made of a permanent character, except on the Monarch and Buffalo claims."

By now, operators had spent years studying the lodes in the Atlanta area. They had learned a great deal about the typical composition. The silver in the ores was actually worth more than the gold. In fact,

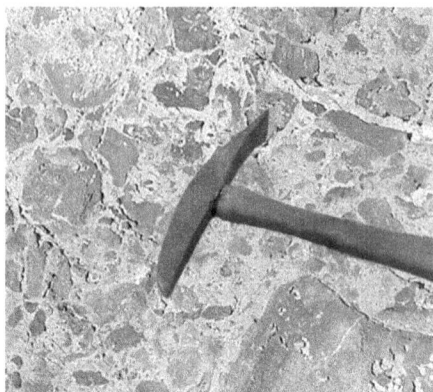
Hand-forged Sampling Pick

the *Avalanche* article said, "The free gold constitutes 20 to 40 per cent of the values."

Owners had also found that the silver occurred in several chemical forms. Yet that knowledge did them little good because they did no know how to efficiently process such difficult materials.

The huge potential returns led some owners to finally bring in professional help. One such was Captain James Baxter, a native of England and a Civil War veteran. Born in 1834, he was an infant when the family emigrated to the United States. Baxter had earned a degree in mining and metallurgy from Columbia College (now University). He also had practical experience as a master mechanic.

When the Civil War broke out, he entered the Army as a private. More educated than most, he soon rose to the rank of Captain. Wounded several times, Baxter spent

Engineer James Baxter

two years on crutches after the War. When he recovered, he ran mines, smelters, and railroads in South America and Mexico. He also gained experience in the West. In 1874, he took over management of several mines in the Atlanta area. The results were encouraging. The *Idaho Statesman* reported (June 15, 1876), "Capt. James Baxter has commenced crushing custom ore from the 'Brick Pomroy' and 'Last Chance' ledges – will soon commence on a rich body of 'Leonora' rock."

Baxter left Idaho in 1882 for work in other U. S. mining districts and in Mexico. He returned ten years later to open a very successful foundry and machine shop in Boise City.

Owners in Atlanta really needed a better wagon road from Rocky Bar to make it possible to upgrade their equipment and processes. Success came only after many abortive attempts. But finally in May 1877 the Atlanta Wagon Road Company opened bids for construction work. It took over a year, but was finally done and said to be "in

Teams Hauling a Boiler Up an Idaho Grade

excellent condition." The *Idaho Statesman* said (August 30, 1878), "As an evidence of the lightness of the grades and the capabilities of the road, the Buffalo company ... transported over the road a huge boiler for their new quartz mill."

The article also said, "The successful prosecution of this enterprise and its early completion reflects great credit upon Capt. Bledsoe, under whose personal supervision the work was done."

That Captain was, in fact, the man who led the second large expedition into the Boise Basin in October 1862 (Chapter Two). Born west of Louisville, Kentucky in 1832, Relf Bledsoe traveled through Mexico to California in 1850. A few years later, he made his name as an Indian fighter in the Rogue River War, in Oregon. The Idaho gold rush drew him first to Orofino and then Florence, where he owned a store.

Captain Bledsoe

After helping found Placerville, he managed a mine in Rocky Bar. Bledsoe never struck it rich in the mines. For a time, he served as an express messenger – a very dangerous job – for Wells, Fargo & Company. He also chased road agents for the company and then had a job as an under-sheriff in Boise City.

After managing a mine near Silver City, he returned to run a mine near Atlanta. It was then that he took the contract to build the Atlanta-Rocky Bar road. Within a few years, he followed the rush into the Wood River mines.

45

Buffalo Mill, Near Atlanta

The tremendous quality of their ore kept the Atlanta mines going for over a decade. Thus, a correspondent for the *Idaho Statesman* reported (June 20, 1885) that several mines were still doing well. Referring to the Atlanta Mining Company, he wrote, "They have a large quantity of first-class ore ready for milling, and a large quantity of what they term second-class, but which in other camps would be looked upon as first class … "

However, within just a year or two, the best ore ran out. Operators found that the "second class" ore was not quite as good as they had hoped, and the recovery processes were simply not up to the job. A large sale of properties to some English investors in 1891 spurred hope. But they too gave up after spending several hundred thousand dollars on development.

Another infusion of new cash in 1894 kept people working at the Atlanta mines for a number of years. That encouraged other ventures. The *Idaho Statesman* echoed (August 28, 1895) the optimism of the locals: "Various other operators around Atlanta are doing well and the town and whole region is again booming."

"Booming" is, of course, a relative term. Because of the continuing difficulties in processing the unusual ores, profits were sparse to non-existent. The so-called boom around Atlanta would last only three or four years.

The mines around Rocky Bar also continued to produce steadily into the mid-1880s. By careful management of their properties, some operators actually did fairly well.

Among those was Thomas L. Johnston, who had entered the Basin with Bledsoe and then moved on. In 1876, he and a partner went East to look for development capital. The *Owyhee Avalanche* said (May 20, 1876), "They report that several Eastern men will be out here this season for the purpose of investing in our mines."

According to historian H. T. French, Johnston "made and lost a number of fortunes," largely because "dishonest partners were the bane of his life." That latter point seems to have been accurate, given the litigation he found himself involved in, especially in 1880-1881. He perhaps tired of all that after awhile. In December 1878, at Rocky Bar, he had married school teacher Mary Elizabeth Craker. She had brought her sick brother West, where she nursed him back to health.

Dr. Mary E. (Johnston) Donaldson

When Mary expressed a desire to attend medical school – almost unheard of at that time – Thomas agreed. In 1892, she graduated from Wooster University, in Cleveland. Johnston then financed a series of hospital/health spas in Oregon.

They moved to Boise City and built a "sanitarium" there in 1896. Sadly, Thomas passed away the following year. Captain Relf Bledsoe was one of the pallbearers at his funeral service.

As the more-accessible high-grade ores around Rocky Bar began to play out, operations entered a relative lull, similar to Atlanta. A 50-stamp mill built in 1886 by new investors from England had only about three years of profitable operation. After that, it went further and further in the red.

50-Stamp Mill Near Rocky Bar

In 1888, the *Idaho Statesman* reported (May 20), "The Alturas company paid off their men yesterday. ... They have only about ninety men employed now, forty less than usual and run but thirty stamps." Ever optimistic – a necessary trait in the mining business – locals felt that new financing would soon get things going again. But there, as in Atlanta, several years passed before they experienced another flurry of activity. Moreover, it is generally conceded that extensive and profitable lode mining around Rocky Bar ended in about 1892.

A disastrous fire, which started in the Alturas Hotel, surely contributed to the decline. Writing about the conflagration, the *Idaho Statesman* claimed (September 2, 1892), "A great fire occurred in Rocky Bar, Elmore county today. The town is almost totally obliterated."

Alturas Hotel, Where the Disastrous Fire Started

The town was rebuilt and would do all right for another twenty-five years. However, despite some expansion attempts, the nearby mines never again became major producers. Rocky Bar survived basically on the remaining small-scale operations, and as a supply depot for Atlanta.

On the other side of the Basin, the lode miners around Quartzburg had perhaps better reasons for optimism in the years leading into the mid-1890s. Ominously, however, they were largely dependent upon one main resource. The *Idaho Statesman* said (June 22, 1886), "Quartzburg is the location of the Gold Hill Mining Company's property. This company is what makes the town, and without it the town would soon be a deserted village."

William Coughanour, who had managed the Gold Hill in the Seventies, moved to Payette in 1885. However, he would hold an interest in the mine for over twenty years. His Uncle David Coughanour continued to oversee the property.

David Coughanour Home in Quartzburg

In fact, in 1874-1875, David Coughanour built a fine home in Quartzburg and lived there until his death in 1904. In Payette, William invested in a lumber operation and a bank, as well as several farm-ranch properties. Coughanour was one of the wealthiest, most influential leaders in Payette when he passed away in 1936.

A few other properties seemed to show promise, but had not yet been proven. One of those belonged to Jonas W. Brown, who had come to the Basin in 1862 (Chapter Two). Although for many years he lived and practiced law in Idaho City, he owned interests in several mines in other areas. Like Coughanour, he left the Basin, moving to Boise City in 1882.

Quartzburg, ca 1892

Placerville, ca 1892

During this period, Placerville had its ups and downs as prospectors found good quartz pockets, and then the ore played out. Lacking a reliable long-season water supply, few owners pursued placer mining on a regular basis.

On the other hand, placers continued to produce well along Grimes Creek – Centerville and Pioneerville. Pioneerville was doing the best. In a review of Basin mining, the *Idaho Statesman* listed (September 28, 1890) several good properties. That included one where, "They are obtaining a paying quantity of ore at a small cost."

Prospectors also found good lodes east of Pioneerville. Noting the production of one of these sites, the *Statesman* said (December 12, 1895), "The successful working of this mine will cause money to be invested in other properties in the vicinity."

Across the dividing ridge on Mores Creek, miners had also done well with both placer and lode mining. The *Statesman* published (June 19, 1886) "an interesting letter" from Idaho City that said, "Placer mining has dwindled down from its former immense proportions, but in the aggregate a good deal of gold is annually washed out of the gulches and off the hill sides."

Hotel Man Galbreaith

The writer also said, "The quartz mines of Boise County are looking well and promise good results when properly developed."

Through the Eighties and into the Nineties, these kinds of results kept the merchants in Idaho City going. Occasional "boomlets" from rich finds raised hopes for even better times ahead. One business that flourished was the Luna House hotel, managed by Walter S. Galbreaith. (Records show at least three variations on the spelling, including "Galbreath" and "Galbraith.")

Born in Shasta County, California in 1861, Walter lost his father in a hunting accident four years later. His father had come to the Basin in 1863 and seems to have done very well. Because of the great 1867 fire, the records are somewhat muddled, but the father apparently owned considerable property in the area.

Walter's mother remarried in early 1867. The stepfather, Matthew G. Luney, brought the family to Idaho later that year. Both Walter and his sister Clara had scary memories of riding the stagecoach from Winnemucca to Silver City during the "Snake Indian War," which engulfed southwest Idaho at that time.

Also that year, what came to be called the "Galbraith House" was built in Idaho City. It's not clear if the house was a replacement for one destroyed in the 1867 fire. In any case, the home belonged to Galbreaith descendants "well into the 20th century." Today, the Galbraith House is considered the oldest residential structure in Idaho City.

Galbreaith House, 20th Century

Walter first tried his hand at mining, when he was about twenty years old. That venture was not very successful and he ended up working for his stepfather at Luna House. He eventually became a manager and then sole proprietor of the hotel.

In 1887, Idaho City lost one of its most prominent businessmen, brewer Nicholas Haug. The death notice in the *Statesman* (July 26, 1887) said "He has been in poor health for many years and his death was not unexpected. ... The remains will be taken to Idaho City to-day where the funeral and burial services will take place."

Masonic Temple, Idaho City, Built 1867

Since Haug was a member of the Masonic Order, the Idaho City Lodge no doubt held an appropriate ceremony in their hall. The Masons had been active in the Basin from the very first. In 1865, James Pinney – merchant and theater buff – led the effort to build the hall, said to be "one of the finest in the Territory."

The Temple – perhaps the oldest west of the Mississippi – still stands today, although it did lose the free-standing bell tower to a violent wind storm in 1958.

The Basin experienced something of a down year in 1888. Capital was apparently scarce for new development, and the reliable Gold Hill Mine had been shut down for a program of repairs. As the *Idaho Statesman* said (July 22, 1888), "This is an off year for Idaho in her production of bullion."

The following year proved somewhat better. However, Governor Shoup's repeat call for a constitutional convention created the most excitement in Idaho Territory. While the Boise River gold country had its up and downs after 1870, mining – mostly involving lead-silver lodes – boomed in other areas of the state. Moreover, stock raising and farming had become major components of the economy. The population had blossomed almost six-fold since the 1870 census.

Shoup's call was not technically legal since Congress had not passed any "enabling legislation" to form a proposed state structure. But as James H. Hawley later wrote, "As a precedent for his action he followed the examples of Arkansas, California, Florida, Iowa, Kentucky, Maine, Michigan, Oregon, Vermont and Wisconsin, all of which states adopted constitutions and secured their admission into the Union without enabling acts."

Governor, Later Senator Shoup

Perhaps a half dozen delegates to the convention had some connection to the Boise gold country. James I. Crutcher, for example, had been Boise County sheriff before moving on to Silver City. Born in Shelby County, Kentucky, just east of Louisville, Crutcher followed the gold rush to Colorado in 1860, when he was twenty-five.

He moved on to Idaho in 1862, and then to the Boise Basin the following year. Crutcher did moderately well in the gold fields, but in 1865 he found himself embroiled in trouble. He had an appointment as Under-sheriff of Boise County when the Patterson-Pinkham shootout occurred (previous Chapter). After the initial action, the Sheriff decided he needed no more excitement, and resigned. Commissioners quickly elevated Crutcher to that position. He performed so well that voters elected him for the job in the next election.

Sheriff Crutcher

At the end of his four-year term, he went back to mining. By 1871, he owned a partnership in a Boise City saloon and was involved with mining investments in and out of state. When the governor called for the convention, Crutcher was known well enough around Silver City to be selected to represent Owyhee County. He would later be appointed U. S. Marshal for the state of Idaho. Crutcher passed away, in Boise, in 1915.

Attorney George Ainslie served as the Boise County representative to the constitutional convention. He was born in Boonville, Missouri – midway between St. Louis and Kansas City – in 1838. Like Crutcher, George moved first to

Delegate Ainslie

Colorado before coming to Idaho in 1862. He moved on into the Boise Basin the following year. However, rather than mining himself, Ainslie opened a law practice in Idaho City. (He had been admitted to the bar in Colorado.) Even so, he would be heavily involved in mining investments for most of his life.

From 1869 to 1873, Ainslie had two jobs. He practiced law and also edited the *Idaho World* newspaper. He ended the editorial job shortly before being elected District Attorney for the region that included Boise County and nearby jurisdictions. Starting in 1878, he served two terms as Idaho Delegate to Congress, but was defeated in his bid for a third term. Partly due to shaky health, in 1890 George relocated to Boise City, where he bought an estate across from the U. S. Assay office.

Celebrating Statehood and Independence Day in Quartzburg

The constitutional convention completed its work on August 6, 1889. Idaho became the 43rd state in the union on July 3, 1890. The new status added special zest to Independence Day celebrations held all over the state.

Miner and Hardware Man Kennaly

Hardware dealer John Kennaly, Sr. was among those who celebrated the occasion in Idaho City. He had been born near the Canadian side of Niagara Falls in 1833. He arrived in Idaho around 1862 and, two years later, partnered in a Boise City hardware store. After a year or so, he sold his interest and headed east to visit his parents, who had moved to Ohio.

In 1866, Kennaly returned to Idaho, where he established a base in Idaho City. John made his home there for the rest of his life. Records show that, in 1868, he was appointed Grand Marshal of the Idaho City Masonic Lodge.

He may have owned interests in several mines, but his principal mining activity was well north of the Basin. In 1877, he partnered with long-time Boise City pioneer Frank R. Coffin in an Idaho City hardware and implement business. According to historian Hiram T. French, their firm "developed into the most important enterprise of its kind in the state."

In 1891, Kennaly, along with two other members of the Idaho City school board, had to make a decision. The old school that served the area needed to be replaced. They tried to sell the old building, but received no bids. They finally tore it down and sold the materials piecemeal. Construction began on a new structure in June. The *Idaho Statesman* reported (June 19, 1891), "The work of cutting the stone for the foundation of the new school house here commenced yesterday. The carpenters will begin framing the timbers in three or four days."

Unfortunately, their funds ran out before the building could be completed. Kennaly and the other trustees volunteered their labor, and a community dance brought in about $140. Completed in September 1891, the building housed the local school for over seventy years. Today, Idaho City still uses the structure as its city hall.

1891 Schoolhouse. Now City Hall for Idaho City

Going on into the Nineties, it was "business as usual" for the Boise River mines. The "Panic of '93" – a worldwide economic depression – brought a shortage of development capital, but Basin operators were inured to that. (The Panic devastated silver producers in the Coeur d'Alenes.) Sadly, it did force John Kennaly out of the hardware business; he had to give up his entire inventory to satisfy creditors.

By an odd coincidence, the same June 1891 newspaper item that covered the schoolhouse foundation work also said, "John Kennaly, of this place, who has fallen heir to a large fortune, left for New York City last Sunday morning to be identified."

Now in his sixties, Judge Kennaly – he had been elected four times as a Probate Judge – chose to retire on his $100,000 inheritance after losing his business. Not known as a teetotaler, John perhaps spent some time at the local "watering hole," discussing old times and matters of the day with other habitues. His son was appointed Idaho City postmaster in 1911, and John Sr. passed away in 1918.

Genial Crowd in Idaho City Saloon, ca 1900

Saloon patrons would eagerly compare notes about the latest reported finds, who was hiring, who was laying off, what investors were doing, and on and on. Some, like George Ainslie when he dropped in from Boise City, John Kennaly, and other less-known old-timers, could remember when the wealth seemed endless. Before 1870, or even earlier, miners in the Boise River gold country had produced perhaps 1.2-1.5 million ounces of gold, mostly from placer claims. (At today's prices, that would be worth in the neighborhood of two *billion* dollars.)

The region never again come close to that annual production. Surviving mines tended to go from boom to bust, literally from one season to the next. Yet, cumulatively, the placers and quartz lodes had churned out something approaching another million ounces of gold during the span since those heady early years.

The placers had settled into a steady pace, with hydraulic giants gouging deeper and deeper into the high ground. But costs grew as miners had to build longer and more elaborate flume systems to deliver water to their operations.

Hydraulic Giant Washes Hillside Near Idaho City

Prospectors regularly made new quartz finds to fuel hopes for a boom. But no one found, or at least reported, any discoveries big enough to cause a new rush into the Basin towns. In fact, Basin villages were now rather staid, settled communities.

Daly Hotel, Quartzburg

Yet placer mining along the Boise and its tributaries was about to enter a new phase. Huge dredges would dominate the rivers and streams for over a half century.

CHAPTER FIVE

DREDGING AND HARDROCK MINING

Summer thunderstorms in the high country may or may not arrive with some notice. There's a tautness in the air, and the clouds may mutter a warning. But when high ridges crowd close, those portents may not give much time.

The overcast did little to warn citizens of Idaho City of what was about to happen on July 14, 1896. Around 4:00 in the afternoon, lightning bolts began striking all around. Deafening thunder stunned the senses, and torrents of rain pummeled everything, turning the roads into muddy bogs. At John Kennaly's home, a strike blew chunks off a nearby shade tree.

Other bolts ripped into evergreens clinging to the higher ridges. High winds lashed at trees and brush, and drove the rain onto covered sidewalks and into open doors. Frenzied stock dashed against fence and wall. Miss Ida Foster, seated at her kitchen

Lightning Strike Near Idaho City

table, perhaps just escaped violent death from a stray shock. One of her sides remained paralyzed for some time. The *Idaho Statesman* asserted (July 16, 1896) that the town had been "visited by the heaviest thunder storm ever seen here."

After an hour, the clamor moved off into the ranges to the north. About another hour passed, then the heavy rain picked up again. So much rain fell that Mores Creek quickly rose "to a point higher than at any time during the spring."

Despite the shock and damage, most locals rejoiced at the sudden bounty of mid-summer rain. It may well have extended the placer mining season by as much as an extra month, in some places. Still, on August 6, the *Statesman* reported "K. P. Plowman has finished placer mining for the season in his East Hill claim."

Plowman's East Hill Placer

Three weeks later, the newspaper reported (August 28, 1896) from Idaho City: "The placer mining season has about closed around this place, although there are a few claims that will operate until freezing weather."

Most placer miners who stayed at their claims were busy extending or repairing their ditches and flumes to be ready for the next spring. Lode miners, of course, did not necessarily want a lot of rain, since they had to worry about flooding in their shafts.

Both placer and quartz miners saw hope in the country's slow recovery from the depression that followed the Panic of '93. That might free up funds to develop new ground and cut deeper into known ore bodies. In the meantime, owners still had to do most of the work themselves.

However, a transaction in late 1897 signaled a change in the old ways of extracting gold around Idaho City.

Attorney and ex-Delegate George Ainslie, whom we met in the last chapter, was a partner in a firm that held much placer ground near town. They had filed on it in 1889, but could not attract the capital needed to exploit it. Then, in 1893, another company asserted that the prior claim had lapsed, and tried to "jump" the property.

Commenting upon the dispute, the *Idaho Statesman* said (October 17, 1893), "Both sides are firm and a long and interesting lawsuit is bound to follow, if not something of a more serious nature."

Ainslie's company finally retained its rights. They then sold out to a Boston firm after a lively bidding process. The *Statesman* said (November 9, 1897), "The plan is to dredge the entire tract and the machines put in will handle all the dirt that the available water supply will wash."

Partners Working Idaho City Lode Mine by Themselves

After that George Ainslie focused on business in Boise. He moved to California in 1908 and died there five years later.

The ground the dredging company acquired stretched along Mores Creek from about two miles above Idaho City to about four miles below. A side spur ran about a mile up Elk Creek. (Elk Creek joins Mores Creek right near the town.) Compared to what came later, those early dredges were rather modest affairs. A cycling array of buckets

Boston and Idaho Dredge near Idaho City, 1898

grabbed scoops of gravel and mud. At the top of the track, the bucket emptied into a perforated moving table or barrel where a spray of water washed the pay-dirt through.

A conveyor ejected the stones and larger gravel from the separator to the side or well behind the hull. Flat water trays, much like large riffle boxes, then captured the gold.

The shallow draft dredges used in the Basin needed no river. They made their own pond, digging it out at the front, filling it in behind. Such dredges could – and did – gobble up riverbanks and meadows until they hit a wall of bedrock. Quite a few old valley mining camps got sucked into their maws.

Five-Cubic-Yard Dredge Bucket, 1905

The *Statesman* reported (May 23, 1898) an interview with the dredge construction company representative, one D. P. Cameron: "Mr. Cameron said that the dredge … will be ready to run between June 25th and July 1st. The machinery has been somewhat delayed by work for the government at San Francisco."

His company, the Risdon Iron Works, perhaps had contracts involving preparations for the Spanish-American War. A month earlier, Congress had declared a state of war with Spain. The President called for mobilization of the various state volunteer – national guard – units, and Idaho Governor Frank Steunenberg responded immediately

Governor Steunenberg

The First Idaho Regiment quickly filled its rosters. The *Statesman* interviewed (May 4, 1898) two men, one from Placerville, the other from Idaho City, who had come to Boise to enlist. The paper noted, "They say there are many in the Basin ready to tender their services."

It's unclear how many from the Basin finally enlisted. One young man from Idaho City served in the company recruited at the University of Idaho. The First Idaho arrived in Manila after the war against Spain was officially over. They did see action against Filipino guerrillas. The insurgents were fighting for independence

National Guard Troops in the Philippines

from *any* foreign power. The troops returned to Idaho in September 1899.

The U. S. acquired several overseas possessions as a result of the war. That made America a (reluctant) world power at the turn of that century. Meanwhile, Idaho's population had surged over 80 percent since statehood. Yet Idaho City had continued to lose people. The town recorded fewer than 400 inhabitants in the 1900 census.

While placer mining took on a new life with the dredges, other operators sent crews deeper into their quartz lodes. The *Statesman* reported (December 19, 1901) that attorney Harry L. Fisher had leased a mine six miles northeast of Idaho City. It went on, "Mr. Fisher will drift on the hanging wall of the north ledge and hopes, when he gets opposite the big shot in the south ledge, to strike rich ore."

Attorney Harry Fisher

Fisher was born fifty to sixty miles east of St. Joseph's, Missouri, in 1873. He began prospecting in the Basin about 1892. He did not do particularly well, so two years later he enrolled in the Stanford University Law School. He had earlier read law in several law offices. Fisher returned to Idaho in 1896 and was admitted to the bar.

A couple years later, he opened his own law practice in Idaho City. Of course, he never lost his interest in mining, as noted by his quartz mine lease. In 1902, voters elected him Prosecutor for Boise County. He excelled, and easily won re-election to another term. Shortly after his second term expired, Fisher sold his Idaho City office and residence, and moved to Boise.

Besides a fine law career, Fisher also pursued a number of mining and irrigation ventures. For awhile, he owned valuable mines in Lemhi County. Later, he served as President of a mining firm with holdings in northern Idaho and western Montana. He passed away, in Boise, in March 1940.

As predicted, the Boston & Idaho Gold Dredging Company returned substantial profits to its stockholders. In 1903, they brought in well-known mining engineer Forbes Rickard, from Denver, to assess their property. The *Idaho Statesman* reported (October 8, 1903) that, "He will recommend that two and probably three large dredges of the Bucyrus pattern be put in."

Idaho City Dredge. Note Different Design (Tripodal Hoist)

At that time, the company had two dredges in operation. The original 1898 dredge had been steadily profitable. However, a newer and larger dredge had not done so well, mainly due to its less effective design. Rickard also recommended that they use electrically powered units only. The article went on "The Boston & Idaho company will put in an electric plant of its own if it cannot purchase the needed power."

Eventually, several dredges would be running along Mores Creek, both above and below Idaho City. At the same time, prospectors kept finding new quartz lodes. For example, the *Statesman* announced (August 8, 1905) that the location – "between 12 and 15 miles northeast" of Idaho City – of a rich find was now known. "The discovery was made last fall … [But] The secret was jealously guarded until the discoverers were ready."

Although gold production did not draw large numbers of miners into the region, it did bring considerable prosperity to the existing businesses in Idaho City.

A solid business proposition brought newcomer Melvin Wiegel to town in 1905. Born near Sandusky, Ohio, in 1875, Wiegel earned a technical degree from a small college nearby in 1892. He eventually became Chief Engineer for an electric power company in Ohio.

Then a mining company hired him to build a power plant near Idaho City. When the company defaulted, Melvin decided to try his hand at prospecting. He and a partner found a rich quartz ledge, which provided Wiegel the wherewithal to branch into several lines of business. Besides a fine home, a saloon, and other properties, Wiegel owned stock in the Boise Basin Bank, for which he became Vice President.

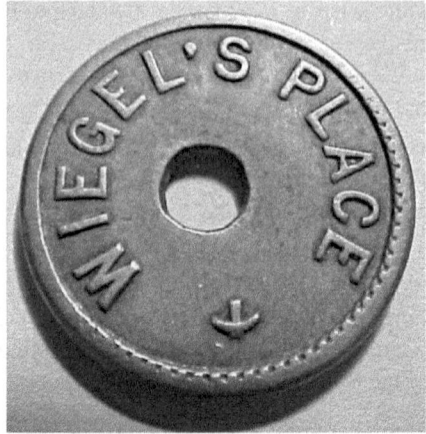

Business Token

An avid hunter – as was his wife – Melvin collected dozens of animal trophies. Somehow, he also became a skilled taxidermist. He displayed those trophies in his place of business for over a generation.

About the same time investors began dredging along Mores Creek, other firms brought similar equipment to the placers along Grimes Creek, and near Placerville. These companies tried to move directly into dredges powered by electricity. Unfortunately, that crippled their enterprise until enough power could be routed into the

Wiegel's Place, Idaho City. Note trophies on the Walls

Placerville Dredge, ca 1899

district. As with hydraulic mining, lack of a dependable water supply meant that a hydroelectric plant on Grimes Creek often had to shut down during the late summer and early fall.

Still, the rich prospects brought many investors into the area. Thus, the *Idaho Statesman* reported (October 28, 1897) that "Montana capitalists" had leased placer ground in Boyle's Gulch, just northeast of Placerville. The article said, "The ground is owned by Myer & Smith, James McDevitt, J. S. F. Bathcen and others."

Born in Ireland, reportedly in Cork, James McDevitt emigrated to the United States in 1863, when he was about twenty years old. He made his way to Idaho at some point and the 1870 census describes him as a butcher and a miner. (He probably spent more time at the butcher shop when the placers shut down for lack of water.)

James McDevitt

Despite the Roman Catholic churches in the Basin, in 1876 James and his bride-to-be, Minnie O'Brien, traveled all the way to San Francisco to be married. A Father Larkin offici-ated, so perhaps they preferred an *Irish* Catholic priest to marry them.

At the time when he leased ground to the Montana capitalists, McDevitt had accumulated a number of mining investments. According to the *Statesman* (October 27, 1894), he had also been appointed as postmaster in Placerville. He ran the post office out of the same structure where he had his butcher shop.

In 1899, the *Statesman* headlined (August 18, 1899), "Fire Sweeps Placerville from Face of the Earth." Because the fires raged through the residential section, the article said, "A number of people are destitute in Placerville."

As word spread, other Basin towns rushed to organize relief efforts. The *Statesman* correspondent based in Idaho City wrote, "One load of provisions and clothing has already been sent to the sufferers from this place and another will be sent over in the morning."

The article also said, "Among the losses are … McDevitt's butcher shop, in which the post office was located."

Centerville Relief Wagons Headed for Placerville

James and Minnie also lost their home in the fire. They soon rebuilt. Then, less than a year later, the *Statesman* reported (June 22, 1900), "Placerville Again Swept by the Remorseless Flames." The flames took most of the business district, but only one residence. Among those who lost was "James McDevitt, butcher shop and postoffice, loss $5000. He is away and it is not known whether or not he has any insurance. There was no mail lost as far as can be learned."

Butcher Shop – Post Office Built by McDevitt. Structures on Left Burned Down Later

McDevitt built a new structure, this time using brick and insulating the attic with a thick layer of dirt. He ran the meat market out of the left side of the building, with the post office on the right. The structure later went through a series of uses before becoming today's Placerville City Hall. It now also houses a small store operated by Skip and Donna Myers. McDevitt continued to hold and lease mining properties in the area for a number of years. He passed away in April 1919, less than six month after Minnie, and is buried in the historic Placerville Cemetery.

The Honorable James Henry Hawley

Investors were also active in the quartz mines north and west of Placerville. The *Idaho Statesman*, reported (September 17, 1903) that James H. Hawley and some partners had bought "a promising mining property at [the] head of Canyon Creek, four miles west of Placerville. … The new owners have outlined an extensive plan of development for the property."

Judge Jonas Brown

Hawley had stayed in Idaho City about ten years before moving to Hailey. Then, in 1890, he established his practice in Boise. There he handled many famous legal cases.

But he retained his interest in mining. In 1897, he chaired the Board of Directors of the Boise Mining Exchange. The Exchange specialized in mineral-related investment transactions and news. Surely not coincidentally, Jonas W. Brown was on the Rules Committee. By then Hawley and Brown had known each for probably thirty years.

Brown served as a Probate Judge from 1903 to 1907. "Uncle Jonas" could have continued in the office, but at age 82, he chose not to. He died from a sudden heart attack in September 1916.

At the time Hawley invested in the Canyon Creek property, he was also serving as Boise Mayor. He later served a term as Idaho Governor and ran for Idaho's U. S. Senate seat. (He ran twice, but was defeated both times.) He passed away in 1929.

Two years after Hawley made his Canyon Creek investment, the *Statesman* published (May 21, 1905) a glowing report from the State Mine Inspector about the Iowa Mine, near Quartzburg. The Inspector said that "the natural conditions are exceptionally favorable to rapid and economical operation."

Quartzburg, 1900-1905

He also said, "The Belshazzar is probably the best developed body of free milling ore in the whole Basin country."

The Belshazzar Mine is located less than two miles southwest of Quartzburg. Investors incorporated a company to exploit the mine in July 1906. Although the property shipped some gold over the next few years, it shut down in 1909. The company's charter lapsed three years later.

Meanwhile, over on the other side of Boise River gold country, mines around Rocky Bar had been providing tolerable returns. Thus the *Idaho Statesman* reported (February 3, 1897) that " the New Nicholas is running steadily and employs a force of from six to eight men. The Vishnu mine has just cut a chute of exceptionally good ore while the Elmore company are [*sic*] doing extensive work on their property ... "

Rocky Bar, Street Scene

Six years later, the paper published (May 13, 1903) an interview with a mine owner from the town: "The Vishnu, he said, was at work with a 12-stamp mill and was making big clean-ups." As for the Elmore, "He [had] heard recently that it was being sold to Colorado people, who would operate it."

The mines around Atlanta continued to show much promise. However, the lack of an all-weather road still hampered full development of the rich ores there. For years, there had been talk of a road along the Middle Fork itself. Although some deep cuts would be needed where the canyon narrowed, advocates felt those same high ridges would protect a road from winter closures. But basically, in the early 1900s, most operators around Atlanta were just hanging on.

Town of Atlanta Prospering

Still, in an assessment of the "Mines of Atlanta District," the *Statesman* said (October 6, 1907), "There are millions of tons of ore of commercial value blocked out in the mines of the Atlanta district ... [with] hundreds of good properties."

Three years later, owners were realizing some of that promise. The *Statesman* echoed (October 10, 1910) a claim that "the mining town of Atlanta is experiencing the most prosperous period in its history." According to an observer they interviewed, "The town of Atlanta is booming, and is more prosperous than ever before. Over 200 miners are employed in and around the town."

Encouraged, prospectors pushed deeper and deeper into the ranges east and south of Atlanta. Commenting on one new district that was "looking well," the *Statesmen* said (September 11, 1911) "Work for the season has now started on the Red Horse and May Day groups, owned by R. P. Chattin & Company."

Robert P. Chattin was born in 1863, fifty to sixty miles southwest of Knoxville, Tennessee. He visited the Rocky Bar region for a few months in 1884, then returned to Tennessee to get married. The couple returned to Idaho in May 1886. After periods in Mountain Home, Pine, Glenns Ferry, and Rocky Bar, they finally settled in Mountain Home in 1895.

Throughout this time, and after, Chattin prospected himself, and also studied other properties. In 1902, he bought what was called the Franklin Mine near Pine, Idaho. Pine is about ten miles down the South Fork from Featherville. Fur trader Alexander Ross, noted in Chapter One, passed right through the area in 1824.

Mining Investor Chattin

The Franklin Mine had been a modest, but steady producer since its discovery in May 1887. Two year after Chattin bought the mine, the *Statesman* published (July 15, 1904) an interview with him. His lease-holders had run their tunnel out of one ore body "through some barren ground" and into another pay ledge. The article said that "The vein seems larger but the ore bodies are about the same size and value."

The mine continued to produce for perhaps ten years, but was shut down by about 1914. By then, Chattin had apparently sold the property. As noted above, Chattin owned other mining properties, including the Red Horse, which was about nine miles southeast of Atlanta. His returns allowed him to invest in several banks and much real estate. Unfortunately, he passed away in early 1920 in California, having gone there for medical treatment.

Stagecoach Leaving Pine, ca 1908

Despite the optimism of October 1910, a year later the *Statesman* headlined (October 12, 1911), "Miners Leave Atlanta." By then, companies had suspended operations at every big mine in the area. No one knew why, and no one had heard any date for a restart. The article noted that "75 miners left Atlanta recently within one or two days." Locals could not understand the sudden shut down, since vast bodies of rich quartz were known to remain.

The problem was the difficult geochemical makeup of that ore. Companies still lacked an economical process to recover gold or silver from the matrix. Even so, as the locals hoped, the rich ledges kept drawing new money to the district. Two years later the *Statesman* reported (August 3, 1913) a new find at one of the existing mines. The article said, "The ore comes from a grade deeper than any present workings of the camp, and is of a character that causes mining men to believe Atlanta is due a re-awakening."

"Mining men" often made such pronouncements, of course, which came true considerably less often. In this case, they turned out to be right. The following spring, the

Bagdad-Chase Mine, near Atlanta

Statesman interviewed the Superintendent of the Bagdad-Chase Mine at Atlanta. The article (March 21, 1914) said, "A four-foot vein was struck last fall in the Bagdad-Chase mine, running at from $50 to $200 per ton. The strike revived confidence in the property, and as soon as spring opens the company plans to install considerable new machinery and largely increase the working force."

Two years later, the outlook continued to look promising. The *Statesman* published (October 1, 1916) a long article by Robert Bell, State Mine Inspector. Bell wrote, "The future of the property at this time, and of the Atlanta lode in general, looks more rosy with the promises of profitable results, permanent and extensive mining operations, than it has at any time since its original discovery."

Born in Yorkshire, England, Robert Norman Bell emigrated to the United States in 1881, when he was just sixteen or seventeen. He worked on a Wisconsin farm for a year or so, then traveled west to Montana. There, he worked on the railroad and then as a coal miner. In 1884, he began prospecting and mining in the gold fields north and west of Salmon City, Idaho.

To better himself, Bell studied tracts from the U. S. Geological Survey and subscribed to various mining industry trade journals. Then, in 1891, a publication called the *Colliery Engineer and Metal Miner* began to publish a "correspondence column." That quickly segued into an actual "online" (U. S. Postal Service) "school of mines." Demand for their courses exploded. Within three years, the formally incorporated

Mine Inspector Bell

"International Correspondence School of Scranton, Pennsylvania" had eager students all over the country. Robert Bell was one of them.

Even before that, apparently, Bell had published trade journal articles based on his practical observations. With his new education, he began to publish even more widely. Soon, mine managers and investors sought out his expertise. At the time, State Mine Inspector was an elective office, and Bell first gained that position in 1902. After three two-years terms, he chose not to run the next time. However, he ran again in 1910, and was then re-elected several more times, with larger and larger majorities.

Bell remained optimistic about the industry, but America's entry into World War I changed the environment. A few months after Congress declared war, the *Statesman* headlined (July 27, 1917) the closure of one of the major mines near Atlanta: "Manager Leo J. Falk, speaking of the closing of the mill, said that labor has been extremely hard to get and supplies have been high in price, and the company had thought best to close the mill for a time and spend the rest of the summer doing developing work."

Mill that Processed Ore From Gold Hill and Other Lodes

Quartz mines in the Basin itself went through much the same ups and downs as the Atlanta and Rocky Bar mines. Even the old reliable Gold Hill Mine went through a hiatus of several years before being revived in 1909. The *Idaho Statesman* reprinted (August 2, 1909) an article from the *Idaho World* about the re-opening. It said, "The shaft of the Gold Hill will be sunk to greater depth, and a great amount of ore opened in the mine that could not be worked in the early days at a profit."

Fortunately, less that two miles south of the Gold Hill, dredge operations had been making steady profits for a number of years. The *Statesman* noted (July 25, 1910) that, "The shortage of water, affecting so many, does not in the least interfere with its work as it is so arranged that when the supply gets too low it can be pumped back into the pond and used over."

Placer and lode mining in the Quartzburg-Placerville area held reasonably steady for much of the decade. However, such operations became more and more problematic. The *Statesman* published (June 6, 1919) some findings based on the observations by Mine Inspector Bell. The item said that the Gold Hill was "making better annual yield of bullion than ever before in its history, although the profit end of the operation is cut to the bone, due to the present excessive cost of labor and material."

It is perhaps significant that Bell chose not to run for re-election in 1920 or thereafter. He pursued private interests before passing away in 1935, in Boise.

Hydraulic Giant Strips Gravel Below Pioneerville

Along Grimes Creek, further south of Placerville, returns began to dwindle somewhat sooner. Around the turn of the century, hydraulic miners attacked every bit of ground they hoped would be productive. As shown in the photo, that included pushing toward the bench under Pioneerville itself.

But even with such measures, placer mining near Pioneerville and Centerville seemed to have played out around 1910. Valley placers get little mention in the newspapers after then. In 1915, the Boston & Idaho Gold Dredging Company, which owned dredges on Mores Creek, did prospect placer ground above Pioneerville. However, nothing seems to have come of that. Some of the quartz mines in the higher ground apparently did produce decent returns for small operators. However, it's not clear if even the minor activity there survived the manpower and cost pinch of World War I.

As a sign of how times were changing, the *Statesman*, headlined (November 9, 1915), "Automobiles in Collision." The brief dispatch reported that two motor vehicles, one driven by a John McDevitt, had collided on the road between

Sporty Roadster, ca 1915

Boston & Idaho Dredge Near Idaho City

Centerville and Idaho City. The article said, "They apparently misunderstood as to which had the right of way. The McDevitt car continued to Centerville, somewhat battered in front." The other vehicle had to be towed into Boise for repairs.

Dredge operations around Idaho City remained productive much longer than anywhere in Boise River gold country except Placerville. Much of the time, Boston & Idaho had a least two dredges working. The *Idaho Statesman* commented (April 1, 1912) on the company's larger dredge: "The average life of a dredge like this one is said to be seven years and it is the opinion of the operating company that enough dirt is available to keep the big machine in operation that long."

However, with improvements in the technology, the company ceased routine dredge operations along Mores Creek around two years early, in 1917. They did run it now and then after that.

Fortunately, several of the quartz mines in the area remained productive. Moreover, prospectors continued to find new ledges that looked promising. Of course, mines in the Idaho City area also suffered from the impact of World War I. Thus, on-going newspaper advertisements in late 1917 listed several openings at the Lucky Boy Mine, about five miles northeast of town.

World War I Recruiting Poster

75

Luna House, Idaho City, 1895 Engraving

All that activity kept Idaho City going. By now, Walter Galbreaith and his wife had been managing the Luna House hotel for over twenty years. For a time, Walter served on the Idaho City Board of Trustees, along with Melvin Wiegel. He helped organize the Boise Basin Bank, and became its President. He also owned much Idaho City real estate and invested in many mining claims around the region.

In 1920, Boise County voters elected Galbreaith to the state House of Representatives. Since the legislature convened at the height of winter, Walter and his wife closed the hotel for the duration. Thus, an Idaho City correspondent reported (December 19, 1920) to the *Statesman* that "Mr. and Mrs. Walter Galbreaith left Thursday for Boise, where they have rooms at the Overland. They will stay till after the legislature meets."

World War I severely impacted Boise River gold mining. Yet overall gold production for the state actually increased from 1917 to 1918 … by nearly 8 percent. However, most of that increase came from other regions. Boise Basin dredge output showed a marked decrease. A post-war recession hampered recovery because it limited the investment capital available to develop quartz lodes or expand prospecting for new placer ground. The gold country would mark time for several years.

Luna House Lobby

CHAPTER SIX

BIG TIMBER TAKES AN INTEREST

Balancing carefully on the timber frame, the top sawyer guided the whipsaw along their line. The bottom man hauled on the double handle, leaning hard to force the sharp teeth through the green log. Grrrrr … thump. A wind gust blew sawdust down his shirt and into his face. He spat, then cursed softly. The top man drew the saw back and set its bite solidly.

After three more cuts, the bottom sawyer paused while the other pounded a wooden wedge into the cut so the springy log didn't bind their blade. Despite the winter chill, sweat drenched the bottom sawyer, causing the wood dust to cling to his skin.

Whipsawing Planks is a Two-Man Job

A few minutes later, the two swapped positions. Steadily, they worked down the log, creating a plank ten to fifteen feet long. Fading light finally forced them to stop. As they built up their fire to cook supper, they counted the day's work. They had added about a dozen planks to the growing pile. Soon, the two could load up their wagon and head into Idaho City.

Remnants of a Miner's Log Cabin

Chapter Two contains a brief mention of the fact that the first (and only) county seat of Boise County was the "log cabin town" that became Idaho City. The miners had arrived late in the year and did not want to try to over-winter in tents. But cutting planks is slow, arduous labor, so they felled trees and threw together about twenty log cabins.

Pennsylvanian Peter Pence was among the men who built winter quarters. Born 25-30 miles northeast of Pittsburgh, Pence moved to Kansas in 1857, when he was about twenty years old. For the next four years, he bull-whacked for a freight outfit. Three trips to Denver sparked his interest in the West, and he drove an emigrant team to Oregon in 1861.

He and a partner followed the rush to Placerville and then on to Idaho City. Pence tried prospecting for awhile, but then he and his

Sawyer and Cattleman Peter Pence

partner found they could do better – splitting about $25 a day – whipsawing lumber for sluice boxes, flumes, and other structures. They cleaned up at the end, selling all they produced at a $30 rate to two men who were anxious to build a saloon. After some ups and downs, Pence became a pioneer cattleman along the Payette River. He was a very wealthy man when he passed away in 1922.

By the spring of 1863, entrepreneurs had seized the opportunity and built two saw-mills along Grimes Creek, one near Centerville. Both were water powered. However, by mid-summer, one "Major Taylor" had built a steam-powered sawmill near Idaho City. Both water- and steam-powered mills would soon proliferate in the region.

Water-Powered Sawmill in Deep Forest

The first of the Grimes Creek sawmills was built by Benjamin L. Warriner. Warri-ner was born in northwestern Massachusetts, a major logging area at the time. In 1849, at the age of twenty-nine, he traveled around Cape Horn to California. He may well have made the trip as a crewman: The Port of Bath in Maine issued him a Seaman's Protection Certificate, basically a proof of American citizenship, in December 1849.

If he did go as crew and then jumped ship, he had plenty of company. Early in the gold rush, over two hundred abandoned ships crowded San Francisco Bay. War-riner spent over a decade in Trinity County (northern California), which was both gold country and a source of lumber.

From there, he moved to Boise Basin in 1863 and dealt in lumber there, until about 1875. The following year Warriner relocated to Emmett, and was also elected for one term in the Territorial House of Representatives. He remained there only a few years, and by 1880 was back in the lumber industry at Buena Vista, near Idaho City. Warriner lived there until his death in June 1889.

Idaho City, Buena Vista Bar at Top Right, ca 1885

With all the new sawmills, the price of lumber dropped. Thus, when Fathers Mesplie and Poulin built the first St. Joseph's Catholic church in the fall of 1863 (Chapter Three), they paid about $100 per thousand feet. Although that was still considered fairly high, it was down dramatically from the earlier $250-300 per thousand.

Over in Rocky Bar, the *Idaho Statesman* reported (August 16, 1864), "[The] Comstock, Cartee & Co. steam saw mill is turning out forty-five hundred feet of lumber a day, yet whip sawing pays, such is the demand for lumber."

Area sawmills did more than supply the miners and settlement builders in the Basin. James Hawley's *History of Idaho* tells us that "Most of the lumber used in the construction of Fort Boise was cut at the [Major] Taylor mill." A similar statement surely also held for structures in Boise City.

During this period, several other men who helped the lumber business grow arrived in the Basin. Among them was Alonzo L. Richardson, a native of Missouri. Born in 1841, 25-30 miles southwest of St. Louis, Alonzo emigrated

Alonzo Richardson

to Oregon with his parents just two years later. After three years apprenticed as a machinist, he followed the rush into the northern Idaho gold fields. Richardson had some success until the summer of 1863, when he moved on to Idaho City.

Although he did some prospecting, including a brief foray into Montana, Alonzo mostly worked at the A. H. Robie Lumber Company in Idaho City.

Early Lumber Yard

According to Bancroft's *History Of Washington, Idaho, and Montana, 1845-1889*, Albert H. Robie was born in Genessee County, New York. However, he told the U. S. Census taker for 1870 that he was born in New Hampshire around 1835. He traveled to Washington Territory in 1853 with a railroad survey team.

Robie arrived early in the new town of Lewiston, where he built sawmills and ran a lumber yard. He was not listed among the earliest settlers in the Boise Basin. However, some time in 1863, he had a sawmill built near Idaho City and established the lumber yard where Alonzo Richardson went to work. Robie moved to Boise City in 1864. (It's not clear when he divested himself of the Lewiston business.)

Richardson rose to manager of the company's Idaho City sawmill in 1866. A year after that, he took over management of another of the firm's sawmills, at which time he moved to Boise City. The sawmill assignment lasted about a year, then he became manager of the Robie & Rossi lumber yard in Boise. (More on Rossi in a moment.) He left the lumber business in 1872.

Moses Hubbard Goodwin also pioneered the Boise River timber industry, and he would stay with it for over forty years. Born in Maine, northwest of Penobscot Bay, in 1834, Goodwin had deep roots. His grandfather fought with Revolutionary War Captain John Paul Jones on the *USS Bonhomme Richard*. After itinerate work as a carpenter in the East and the South, Moses traveled by sea to California in 1861. There, he found a job in a shipyard, building steamers for the river trade.

He too followed the gold rush into Boise Basin in 1863. There, he discovered that miners were over-abundant, but skilled carpenters and builders commanded prime wages. The first quartz reduction mill he helped build was only the second such mill in what would become the state. In 1864, Goodwin supervised the construction of a thirty foot waterwheel, the first of that size ever seen in the region.

The following year, he helped build another mill northeast of Idaho City. That fall, the owners of the first mill he worked on hired him as superintendent for their operation. Goodwin eventually became part owner.

Large Waterwheel, Similar to the One Built by Goodwin

However, a recurrence of an old lung ailment forced him to leave the mountains in about 1872. Goodwin then operated a ranch along the Payette River.

He returned to the lumber business after six years, locating in Boise City. Besides his lumber yard, Goodwin had a sawmill, a planing mill, and considerable timber acreage in the Boise River watershed. For a time, he was reportedly the only manufacturer of doors, sashes, and blinds in the Boise Valley.

Albert Robie's partner, Alexander Rossi, was born along the Rhine River in Germany, in 1828. He immigrated to the U. S. at the age of eighteen. Three years later, he became a Forty-Niner. After several years in California, he moved to Oregon and served as an Army quartermaster in the Rogue River War. At the end of the War, he opened a machine shop in Oregon City, ten miles south of Portland. Unfortunately, a "disastrous" flood ruined that venture.

He moved to Lewiston in 1861 and found a job as supervisor of Robie's sawmills near there. Two years later, Rossi

High-Tech Planer, ca 1875

also moved to Idaho City. There he engaged in the lumber business in some way. He also operated a gold and silver assay office. Some time in late 1866 or early 1867, Rossi moved to Boise City and teamed up with Albert Robie.

As noted earlier, Robie had moved to Boise in 1864. There, he formed a partnership with one C. S. Bush, and the *Idaho Statesman* reported (November 11, 1864), "Messrs. Robie and Bush are about to put up their steam sawmill near or in town, and expect to have the mill running inside of two months."

Not quite a year later, the two dissolved the partnership, with Robie retaining the lumber

Lumberman Alexander Rossi

interests. That business survived a fire that destroyed the sawmill in August 1866. A few months later, Robie and Rossi combined to open a gold and silver assay office in Boise City. However, they closed that business in October and a month later began advertising the lumber business as "Robie & Rossi." That, in fact, became the normal pattern for the Boise River timber business. Most decisions about logging in the mountains would be made in Boise, or at even more distance locations.

Boise City, ca 1866

The firm of Robie & Rossi prospered over the next several years, and was probably the largest lumber operation in the capital city. In 1871, the *Statesman* said (August 5) that their yard was being "abundantly replenished ... " and that "Great stacks of building materials are rising all over they yard, and once in three or four days a fleet of

Ox Teams Ready with a Load of Boise River Lumber

wagons come down from the mountains with immense loads, enough lumber at each trip to build a small town."

Robie began investing in ranch properties in the foothills north of Boise City, and then in southeast Oregon. By 1875, Rossi was basically in overall charge, with Robie as a silent partner. Two years later, Rossi had added another active minority partner at the lumber yard.

Rossi also began cooperating with another Boise City entrepreneur, William B. Morris, on a number of projects. Thus, the *Statesman* reported (July 23, 1878), "Dowling Bros, who have been engaged during the past winter and spring in cutting logs on the North Boise river for Messrs. Morris & Rossi, have succeeded in driving down the river some half million feet of first class logs."

Rossi partnered with Morris briefly before the latter's death in August 1878. Co-

Log Drive: Not for the Faint-Hearted

incidentally, Albert Robie had died just a month earlier. Rossi then bought Albert's interest in the business from his widow. William H. Ridenbaugh, a nephew of William Morris, inherited Morris' interest, and he and Rossi would work together for almost thirty years.

For about a decade after 1884, Rossi and his family lived a couple miles south of Payette. He let Ridenbaugh handle the Boise City lumber business while he exploited timber in the Payette River watershed. Of course, Ridenbaugh had numerous irons in the fire, including real estate, water projects, promoting a branch railroad into Boise, and much more.

By the 1890's, Moses Goodwin, the other long-time pioneer in the timber business, had two lumber yards in Boise City. Beyond his business interests, Goodwin twice served in the Territorial legislature, and would also serve twice as a county Commissioner.

He prospered despite a severe loss suffered one winter. The *Idaho Statesman* reported (February 5, 1890) that an ice jam had broken on Mores Creek and swept eight hundred saw logs belonging to Goodwin down into the Boise River. As the mix of ice and timbers tumbled through Boise City, some brave – or foolhardy – men offered to launch boats to recover what

Ice Jam Blocking Stream

they could. However, the paper said, "Mr. Goodwin would not permit them to take the risks as he said there was so much ice running that no boat could live in the stream."

This disaster – which cost Goodwin several million dollars in product value – was especially ironic. Timbermen generally avoided using ox teams to haul whole logs out of the mountains. Instead, they towed them to the nearest usable waterway and floated them downstream. However, because the previous water year had not provided enough flow in More's Creek, Goodwin had had to store his logs beside it. The last thing he needed was a flash flood.

Skidding Logs in the Forest

In 1895, the *Statesman* enthused (October 16, 1895), "Within a month the Ridenbaugh-Rossi sawmill, one of Boise's largest enterprises, will be running full blast. ... The owners of the mill are W. H. Ridenbaugh and A. Rossi, both men of wide experience in the lumbering business."

Loading Lumber into Rail Car

Alexander Rossi had actually disassembled his Payette sawmill and moved it to Boise. Within two years, all of the "big three" in the southern Idaho timber business showed signs of great prosperity. Ridenbaugh was advertising his lumber yard heavily, and apparently moving a lot of product. A *Statesman* "local brevity" said (March 12, 1897), "Wednesday there was shipped from the Rossi mill a carload of lumber to Ontario and yesterday a carload to Shoshone."

A couple months later the paper reported (June 3, 1897), "M. H. Goodwin is putting up a building on his property on the corner of Tenth and Fort which will be used to cover lumber which he has in stock."

However, nationwide and locally, the logging and lumber industry was entering a new era. In March 1891, Congress had approved the "Forest Reserve Act." The Act authorized the President to set aside tracts of forest as "public reservations." This measure was supposed to stop the rampant over-logging and illegal exploitation of the forests on public lands. By then, untold millions of acres had been stripped in the East and the northern Midwest.

Stretch of Clear-Cut Forest

But confusion quickly arose as to the purpose of these reserves. Were they meant to be totally off limits to *any* utilization, or be open to sustainable activities? The Department of the Interior – which administered the public lands – leaned toward the off-limits "parks" alternative. Moreover, the Act had not defined *how* the federal government would impose the necessary controls, whatever the interpretation.

Soon, two successive presidents had reserved 15-20 million acres. However, because the law basically had no teeth, nothing much changed. Then, in early 1897, President Grover Cleveland, with essentially no public input, added 21 million acres to the reserves. A fire storm of protest blew up.

President Cleveland

To many Americans, especially more recent immigrants, this smacked of the European "royal forests" – which included moors and meadows, by the way. Loosely, these were hunting preserves for the aristocracy, and common people entered, literally, at the risk of life and limb.

A few months later, Congress included an amendment to an appropriations bill that addressed concerns raised by the opposition. The bill was signed into law by newly-elected President William McKinley. The amendment would lead to the creation of a process where lumbermen, settlers, livestock grazers, and other users could harness the forest resources in a controlled and sustainable manner.

President McKinley

The Interior Department struggled to implement its new mandate. They made little progress, so in July 1901 Congress authorized a forestry division within the Department. The new division tried to regularize how they managed the reserves. However, Interior was severely hampered because the Department had no professional foresters on its staff. Desperate, they turned to the Department of Agriculture's Bureau of Forestry for help.

Led in 1901 by experienced professional forester Gifford Pinchot, the Bureau had been in existence for about twenty years. Their experts provided technical advice to timber interests ranging from farmers with a woodlot on the side, all the way up to the giant commercial outfits. They, along with

Forester Gifford Pinchot

the more enlightened timber companies, had watched helplessly as timber pirates of all stripes raped the public forests.

Pinchot welcomed the chance to help, but the cross-Department liaison proved unworkable. In 1905, Congress and the President assigned responsibility for the nation's forest reserves to Agriculture's Bureau of Forestry. A few months later, the Bureau became the U. S. Forest Service. At that time, the reserves contained over sixty million acres of forest.

Two years later, those reserves were renamed national forests. Although Pinchot strongly favored conservation he was also committed to responsible use of the forest. That meant – without then calling it "multiple use" – permitted grazing, timber cutting, mining, and recreation. In some quarters, there was an expectation that permit fees and timber sales would make the Forest Service financially self-sustaining.

Mountain Lumber Camp, Western Idaho

During the early governmental turmoil, the Boise River timber industry more or less proceeded with business as usual. However, they began to feel the impact a few years after passage of the 1897 amendment. The *Idaho Statesman* reported (October 4, 1900) that Alexander Rossi and William Ridenbaugh had found themselves in Federal court "charged with illegally cutting timber on United States land."

The U. S. District Attorney dismissed the case. He did so not because they hadn't cut the timber – Ridenbaugh admitted they had – but because the land was "mineral in character," and therefore not included under the forest reserve law.

Some months later, the government sued Moses Goodwin for over $100 thousand for timber he had allegedly cut from public lands. The *Statesman* said (May 7, 1901), "This is the largest of the series of timber suits commenced by the government against Idaho lumbermen."

Cutting Continued Despite Federal Lawsuits

A couple years later the *Statesman* reported (September 27, 1903), "A verdict was rendered in the United States court yesterday afternoon in favor of the defendants in the suit of the government against A. Rossi & Co."

Rossi had been slapped with a $155 thousand suit "the largest by far sought in any timber suit ever brought in United States court." Despite his win, Rossi had had enough. In April 1904, he and Ridenbaugh sold out to a timber company based in Wisconsin. Both retained their lumber yards, but Rossi passed away suddenly in February 1906.

For some reason, Goodwin's case hung around much longer in the courts. Surely frustrated with changes in the industry, Goodwin sold his major timber business in 1903, although he retain a minority interest. He too kept his own lumber yard. The suit was apparently dropped in late 1910 or early 1911. Goodwin retired at that point, but the following year he too died suddenly.

The sale of the Rossi timberlands and mills was part of a general change in the southern Idaho lumber business. In 1900, various Midwestern investors, with forests playing out in that region, began buying up timber properties in Idaho. This influx resulted in a mind-numbing series of business and legal maneuvers whose details are far beyond the scope of this work. These convolutions developed around two major organizations involving outsiders.

The Barber Lumber Company came into being in the summer of 1902. Several prominent Idahoans invested in the firm, but it was incorporated in the state of Wisconsin and the President was James T. Barber. Born in Massachusetts in 1847, Barber had extensive timber business experience when he moved to Eau Claire, Wisconsin in 1886.

Although he remained President of the Idaho company until at least the end of 1906, he was also then President of a lumber company in Eau Claire. So far as is known, he never made his home in Idaho.

Lumberman James Barber

In late 1902, a few months after the Barber incorporation, Midwestern-based funding led to the creation of the Payette Lumber and Manufacturing Company. Initially, the company focused on timber stands along the Payette River. However, their financial resources would eventually be vital to operations in the Boise River forests.

Iowa-born Edgar M. Hoover became General Manager of Payette Lumber in 1904. He was thirty-eight years old. Edgar then had over twenty years of experience in the lumber industry, the final eleven in Minnesota. He would soon have wide business interests in and around Boise. In fact, Hoover spent the rest of his life in Idaho.

General Manager Edgar Hoover.

Local investors played a role, usually minor, in these two companies. The locally-owned Goodwin properties formed another thread in the tangled web of firms striving to control the Boise timber industry. After a succession of investor and management changes, those interests became the Boise Lumber Company, in 1910. It would survive the shakeout, but succumb to recession in the Twenties.

But the battle continued among the various competitors. "When the smoke cleared," surviving pieces of several companies had merged into the Boise-Payette Lumber Company. The final papers were filed with the state of Idaho in December 1913.

Basin businessmen had wistfully considered the benefits of a railroad into the mountains almost from the time, in 1883, when the Oregon Short Line laid tracks across southern Idaho. Ox and mule trains had too little capacity to economically haul supplies in and products – including lumber – out. Years later, a correspondent from Idaho City to the *Idaho Statesman* wrote (April 17, 1892), "What a fine lumbering business would spring up here if we had a railroad. This whole country is one vast forest of pine and fir of the finest and largest species."

Mule-Drawn Freight Outfit

Then all that outside money began flowing into the Idaho timber industry. Investors saw immediately that they could not expect substantial profits unless they had a logging railroad. Construction of such a line got caught up early in the corporate maelstrom noted above. Some of the resulting skulduggery led to completion of a stub of line to Barber Lumber Company property about five miles southeast of Boise.

Within a year of its formation, Barber Lumber had bought land and started building a sawmill and mill town. Initially called Barberton, the Post Office forced a change

Barberton, Soon to be Barber, ca 1907

to just Barber. The rail connection arrived in March 1905 and the mill shipped its first lumber the following summer. Of course, they had no railroad into the mountains, so their logs had to be floated down the river. The drive made it, but they lost a log driver in the process.

In parallel with these private-sector activities, the new Forest Service had been busy. They had hired people, developed procedures, issued grazing permits, and surveyed the forests that were their responsibility.

Then, on July 1, 1908, President Theodore Roosevelt restructured several tracts to create the Boise National Forest. Initially, existing mining claims – most of the Boise Basin, for practical purposes – were effectively excluded from national forest responsibility. That would change over the years as claims were abandoned or as regulations were interpreted differently. Today, the southern half of Boise National Forest basically encompasses the entire Boise River gold country. In the early years, the new regulations imposed some order on timber selection, without particularly impeding logging activities.

Loggers, of course, needed more than the short stretch of track out to the Barber sawmill. Fortunately, in early 1911 Congress authorized the construction of Arrowrock Dam, about twenty miles up the river from Boise. To meet their construction schedule the U. S. Reclamation Service (today's Bureau of Reclamation) needed a railroad to move vast amounts of supplies to the site.

President Teddy Roosevelt

Log Train Headed Toward Sawmill

But a Barber subsidiary – the Intermountain Railway Company – had made a legal filing on the only workable right-of-way. Thus, with tough negotiating, and perhaps some behind-the-scenes influence, the company cut a deal with the Reclamation Service. Barber provided the right-of-way, and the Service built the railroad. During the dam construction period, Barber leased track usage for their own projects … rail lines into the Boise Basin along Mores and Grimes creeks.

After the merger, the Boise-Payette Lumber Company began laying track in 1914. In May of 1915, they started hauling logs out of the mountains. Two years later, after the Service completed Arrowrock, Intermountain bought their trackage and equipment.

As the tracks pushed deeper into the ranges, the company brought in steam locomotives, generally called "Shays," to handle the loads. ("Shay" was the name of the man, Ephraim Shay, who developed and patented the basic locomotive design.) The Shay delivered tremendous power to the driving wheels, so it could handle massive loads and steep grades. Moreover, that power could be generated by a relatively small engine, so the train could negotiate tight curves around steep ridges.

Logging Country Shay

One confrontation highlighted some of the oddities in how the government handled land claims at that time. Boise-Payette had timber claims along Granite Creek (near Placerville) and upper Grimes Creek (beyond Pioneerville). Naturally, they wanted to run log trains into these stands.

However, an outfit called the Centerville Mining Company owned a placer claim that stretched down Grimes Creek starting about a mile below Centerville. They refused to cede a right-of-way across the property, since that would interfere with their dredge operations. Under "normal" land use guidelines, mineral claims trumped timber, so they thought they had a winning hand.

But the Intermountain had been chartered legally as a "common carrier" railroad company. Under powers granted to such entities, the railroad condemned the necessary strip of land and laid track to the edge of the claim. Then, two miles below (old) Centerville they built a railroad station at what is now New Centerville.

Soon, loading yards peppered the canyons in the Basin. During 1916, the Intermountain hauled logs to produce almost fifty million board feet of lumber. Most of the

Steam-Powered Loading Crane

logs went to the Barber mill. However, just as with the mining industry, World War I brought crippling manpower shortages and inflated prices to big timber. Management coped by consolidating camps and placing them in dense stands that could be worked by reduced crews.

Axemen Chopping Down a Big Tree

Boise-Payette Lumber and the Intermountain did very well during a short-lived post-war building boom. Then a major recession hit the country. Unemployment soared and home-building dropped. Boise-Payette had to lay off close to three quarters of the crew at the Barber mill. That triggered labor unrest that lasted for years.

The companies might have gone into a caretaker mode for a time to keep costs down. Unfortunately, a bark beetle epidemic forced them to send crew into the woods to cut stands before beetles spread into them. Not only could they get the cut before the beetles ruined the trees, they might also create a denuded barrier to further encroachment. In the latter aim, they were apparently successful, but it left the railroad, at least, with virtually no cash reserves.

The economy and the company began to recover in 1924. In a sign of the future, Boise-Payette contracted with a trucking company to haul logs out of one canyon, rather than build another short-use spur rail line. The performance of those early vehicles did not impress management, but that would change.

Trucks Contracted for Hauling Logs, 1924

Despite some recovery, Boise-Payette managed not much more than break-even for the rest of the Twenties and lost money in 1930 through 1932. By now their Intermountain Railway subsidiary was losing money steadily. In 1934, they took another look at using trucks – now greatly improved – instead. Coincidentally, that was the year that most of the Boise Basin became part of the Boise National Forest.

Boise-Payette finally "pulled the plug" on the Intermountain Railway in the spring of the following year. By that fall, much of the roadbed had been converted to a highway. Ironically, a good deal of the traffic on that road consisted of the powerful log trucks that had supplanted the Shays.

Newer Trucks Supplanted the Logging Trains

Over the next few years, the company deployed a larger and larger fleet of trucks, which were proving far more flexible and cost-effective. The *Idaho Statesman* reported (October 9, 1935), "[The] Idaho timber industry has come back from a condition of idle woods camps to near capacity production at bustling sawmills."

The logging and timber products industry continued to grow in importance for the next forty years.

CHAPTER SEVEN

MINING REVISITED

The census taker, William Hooton, continued along Montgomery Street. He had already recorded old Mr. McClintock and Joe Knight, the livery stable operator. Next were the Galbreaiths. He'd known Albert and Sadie for years; their Luna House hotel was older than he was. Of course, he knew most everybody in Idaho City. But the 1920 U. S. Census was serious, official business.

On Main Street, he recorded Melvin Wiegel and his wife Mary. When Idaho went officially dry in January 1916, ahead of country-wide Prohibition, the couple had switched Wiegel's Place to selling soda, cigars, and candy. In no particular hurry, William spread his canvass over four days, being sure to visit the nearby mining sites.

William Wallace Hooton was a native Idahoan, which had become more common early in the Twentieth Century. His father, also William Wallace, had arrived in Lewiston, Idaho in October 1864. From there he had moved on to Idaho City, where he mined, married and raised a family, and then died in 1904.

Idaho City, ca 1920

William, Jr., had mined, worked on a dredge, and owned gold claims himself. In 1920, he was working in the Boise County offices. At various times, he would serve the county as a Deputy Sheriff, Fire Warden, and Commissioner.

The census he recorded showed just how much Idaho City had dwindled since its great days. Including his own family, he found just 140 people in town and the close-in precincts. The Chinese, once a large, thriving populous, had all but disappeared: He enumerated just four.

Mining was still the main occupation, however. Of the 84 residents with identified jobs, 26 were miners, mine supervisors, or mine owners. Only a single forester and four loggers or lumbermen represented the timber industry.

However, mining lagged badly along Mores Creek after World War I. The only dredge producing significant amounts of gold was near Murray, in northern Idaho. Still, the Gold Hill & Iowa mine, at Quartzburg, extracted the most gold of any property in the state. In fact, the *Idaho Statesman* reported (August 15, 1920) that the Murray dredge and the Gold Hill "produced 80 per cent of the gold of Idaho in 1919."

Gold Hill Mine

As the post-war recession eased, hopes rose for renewed production in several Boise River gold regions. The *Twin Falls News* published (November 17, 1921) a long article expressing this optimism: "Gold mining camps … may be expected to respond to the touch of modern mining methods." Moreover, an experienced firm had "just shipped forty carloads of equipment into the field to build two mammoth dredges. Their engineers are said on excellent authority to have estimated 15 years of work ahead of them to clean up their placer field."

And, indeed, the following year the *Idaho Statesman* headlined (June 11, 1922) that mine operations were "Booming in Boise County." They quoted the county sheriff, who claimed, "Most of the mines in the Basin are running full blast." Moreover, two months later one of the "mammoth dredges" started up on the Feather River, about eight miles below Rocky Bar.

Nonetheless, near the end of the year, the *Statesman* reported (December 8, 1922), "The Gold Hill Mining company suspended operations at the property here Saturday, and with the exception of the pump-man, who will be retained to look after the pumps, more than 50 employes were laid off. No … intimation was given as to whether work would be resumed or not."

In fact, it appears that over a decade passed before the mine reopened. Even then, its resurgence would be short-lived.

Two other operations brightened the gold mining picture during the Twenties. First was the Feather River dredge. (Only one was actually built.) The *Statesman* called it a "monster dredge" and gave its dimensions as over 44 feet wide, 105 feet long, and ten feet deep. With superstructure, think of a multi-story luxury home chugging along the river valley.

Feather River Dredge

The dredge had a chain of 67 buckets, each weighing a ton, that could gulp a half-ton of gravel every three seconds. At its maximum extension, the chain head could go down sixty feet to dig gravel. Planners had down-sized from the earlier fifteen-year working projection, however. The *Statesman* said, "It is estimated that about six years will be required to mine the ground for which the dredge was built."

That projection proved to be quite accurate. For three of the five seasons between 1923 and 1927, the dredge produced more gold than any other operation in the state. It was a close second the other two years. The company suspended operation in June 1928.

The other productive operation during the Twenties was at the Belshazzar Mine, located a bit over three miles due west of Placerville. In 1875, placer miners washed several thousand ounces of gold out of the gravel and dirt over-burden. Soon, prospectors found the lode where the placer gold had originated, which included the Belshazzar.

For the next thirty years, the mine would earn praise for ore that "payed splendidly." Yet, for some reason, successive owners performed relatively little development.

In 1896, the *Statesman* reported (July 19) on some samples from the mines in that area: "Another sample that shows gold in large quantities is from the Belshazzar mine." At that time, the owners included David Coughanour, whom we met in Chapter Four, and Walter Galbreaith, operator of the Luna House hotel in Idaho City.

Western Hard Rock Gold Mining, ca 1905

Around 1904-1905, the property began to attract a lot of attention. By then, the mine had three tunnels, one of which ran 850 feet into the ridge. Finally, on July 6, 1906, investors incorporated the Belshazzar Gold Mining Company. The company not only paid to run the mine shafts deeper, they planned for a plant to chemically process the more difficult ore. Whether or not that facility was ever built is unclear. In any case, they closed the mine in 1909, and the company no longer existed at the end of 1912.

A succession of owners developed the Belshazzar sporadically for the next decade. One adit and its cross-tunnels eventually totaled over three thousand feet. (An adit is a tunnel that runs horizontally into the side of a ridge, rather than down a vertical shaft.)

All this development work finally began to pay off in 1926. A history of the mine from the Idaho Geological Survey said that in 1927 the Belshazzar "produced over half the gold production from Boise Basin and was the second largest gold producer in the state."

The following year, when the Feather River dredge suspended operation, the Belshazzar "was the largest producer of gold in Idaho." However, after that the returns steadily declined, and the mine closed in late 1931. For another decade, owners tried to develop new lodes, but World War II finally shut down the operation altogether.

Belshazzar Mine, 1927

Entering the Thirties, some companies saw potential for new chemical processes to return some of the old Atlanta mines to profitability. Word soon got around, and in the recollections of Alma (Nixon) Bartlett, "This was during the Depression and people flocked there. All kinds of people."

Alma's father, originally from Arkansas, arrived in Rocky Bar in August 1900. He later recalled that he was "impressed by the evidence of native gold in the old camp as evidenced by the many rich gold specimens of quartz float displayed in nearly every home and business house in town."

Although, as noted before, the best years of "the old camp" were behind it, Nixon stayed to become a Justice of the Peace. Alma Nixon was born there in 1907. The family later moved to a ranch where she grew up. She recalled getting married to Harold Bartlett "in the summer of 1930." Of course, she went on, "We were working at the time so could only have a weekend honeymoon. We spent that weekend in Atlanta at the Hoffman Hotel."

Hoffman House Hotel, Atlanta

According to her recollections, Harold's job in the mines ran for only a few years. They had to move on in 1935. Other records show that those properties operated through the Thirties, so he must have been involved in the rehabilitation work – cleaning out fallen rock, re-timbering the shafts, and so on.

A memorable event enlivened the gloom of the winter of 1933. A heavy-duty rock crusher in the company's mill broke and shut down the whole process. They had to order a replacement from a factory in the East. It's not clear if the supplier fabricated a new one or had it in stock, but soon the railroad delivered the part to Boise. Then came the tricky part. Freighters still had no all-weather road into Atlanta.

Anxious to get back in production, the firm contracted with a daring aviator to fly the part into the mountains. The pilot cut his fuel supply to the absolute minimum to

Biplane at the Atlanta Airstrip

save weight and took off. It was reported that he had "two more minutes flying time" when the biplane bounced down on the snow-covered runway.

Big companies were not the only people seeking gold during the Depression. Louie LeRoy Packer was born in Ola, Idaho, a tiny hamlet 20-25 miles north of Emmett, Idaho. A skilled carpenter and mechanic, he eventually opened an automobile service and repair station in Middleton. In 1926, he married an elementary school teacher and began raising a family.

Three years later, "Black Tuesday" – the stock market crash – set off the Great Depression. He was then thirty-one years old, with a wife and two little girls to support. Packer soon had to close his shop and join the long lines of men looking for work. Unwilling to take any government hand-outs, Louie searched for an alternative. Then he "met some fellows" who said they had found gold up in the Boise Basin. He told his wife, "They invited me to return with them tomorrow and stake out a mining claim of my own."

Considering it worth a try, they bought a tent and a small cook stove and headed into the mountains. After finding some color, Louie built a riffle box for a more extensive trial run. The results convinced them that the hills still contained enough gold to provide a living.

They then returned to the valley for the winter. By the summer of 1935, Packer had a claim on Spanish Fork, about a mile and a half north of Idaho City. They started with just the tent on the claim, but later built a comfortable cabin. Early on, Louie acquired a partner to help work his holdings.

Campsite Near Packer Mine

Soon, Packer began to improve his claim, rebuilding an abandoned flume system to deliver more water. Eventually, he had a large enough flow to work a hydraulic giant. Louie also prospected for quartz claims and after several years had properties the family considered worth $150 thousand dollars.

Louie Packer and His Giant at Spanish Fork

No one seems to have determined how many small private operations like Packer's there were in the Basin. His daughter's memoir mentions the men who suggested the idea to start with, and "many Chinese."

Of course, Louie's "take" was relatively small compared to what the big outfits recovered. At some point, even the old Gold Hill Mine operated for a time. The State Mine Inspector said that in 1935 it had provided "The largest increase in the output of gold." (However, it shut down again around 1938.)

Besides the lode mines, dredge operations also picked up during the Thirties. These involved units that could dig much deeper, like the one that had worked the Feather River. Older dredges had only gone about thirty feet deep, only half what a modern dredge could do. Best of all, the price of gold increased from around $20 to $35 per ounce. With all that, a big dredge might generate a monthly return of $30 to $50 thousand. Of course, considering the construction cost, they pretty much had to.

Boise County gold production more than doubled from 1930 to 1938, with a four-fold increase in cash value. Most of that came from the Boise River country, including new dredges built on Mores Creek and Grimes Creek in 1936. These began to profitably re-work acres of placer ground.

GOLD DREDGING IN IDAHO CITY, DEC, 2, 1940.

However, that all changed with the advent of World War II. As with the First World War, the mines faced serious manpower shortages and, of course, rationing. Then, early in 1942, President Franklin D. Roosevelt created the War Production Board to oversee rationing and resource allocations.

In October, the Board promulgated Limitation Order L-208. That Order, and various amendments, shut down gold mining as being non-essential to the war effort. Copper and lead mines – which *were* considered essential – needed every man they could get because so many had been drafted.

Anecdotal evidence suggests that the Board's strategy was not entirely successful. If essential-metals mines lay close by, then the gold miners moved over. However, without the allure of the precious metal, many chose less dangerous, though often no less strenuous, lines of work. Some would never come back. Large

President Roosevelt

scale mining did return to the Boise River gold country after the War, but its rejuvenation would be relatively short-lived.

CHAPTER EIGHT

RECREATION AND TOURISM

Captain Griffin watched with satisfaction as one of his companions landed another fat trout. The fish here in Canyon Creek must be starved, the way they attacked the bait. Laughter bubbled from the campsite where the ladies entertained themselves with lazy conversation.

Although the real summer heat had not yet hit the valley, Boise City couldn't match the bracing fresh air here in the Basin. The fragrance of towering evergreens and the impressive mountain vistas made for an ideal setting. And if they needed anything, they could get it from Placerville, only a few miles distant.

Genteel Forest Campsite

They returned to Boise City after a perfect three or four days. The account in the *Idaho Statesman* (June 22, 1872) said the fish swarmed "like so many bees" but "They only caught about 15,000, though it was impossible to keep a strict account."

In 1872, Captain James W. Griffin owned the famed Overland Hotel in Boise City. Born about twenty-five miles north of Bangor, Maine in 1819, Griffin first went to sea in his early teens. He sailed all over the world and became a ship's captain before "the age of manhood." Griffin traded the sea for Idaho in 1863 and purchased an interest in the Overland five years later. A year after that he became sole owner.

Overland Hotel, Boise City, ca 1880

Most main-line stagecoaches stopped at the Overland Hotel. As a result, people all over the northern West knew about it. In fact, "meet me at the Overland" was all the directions anyone needed for a get-together.

An early-day tourist might have taken his (or her) guidance from *Crofutt's Transcontinental Tourists' Guide*. Travel writer George A. Crofutt published his *Guide* regularly in New York City. In his 1872 Edition, Crofutt said the station at Kelton, Utah was "of more importance than any yet passed since leaving Promontory. ... From this station a daily line of coaches leave on arrival of the [railroad] cars for Idaho and Oregon."

He then featured the "Boise Country ... To which the line of stages spoken of convey the adventurous passengers." He also wrote, "The principal mining country is in that portion generally designated as the Boise Basin."

Crofutt did not list any hotels in his *Guide*, but the intrepid traveler would most likely seek accommodations at the Overland, where he got off the stage. After resting the night, he or she could catch the branch stage to Idaho City and other towns in the

Basin. Of course, the Territory had no tourism bureau, so we have no idea how many people availed themselves of this opportunity. Perhaps not many, since Crofutt noted that miners in some parts of Idaho had "experienced considerable annoyance from the Indians, who have been exceedingly hostile."

Idaho City Stage Station

The following year, the *Statesman* published (July 26, 1873) an article that extolled the wonders of the Basin. The writer said, "Our Eastern people boast of Saratoga ... " but "we are not altogether removed from pleasure resorts and the opportunity of breaking the monotony of these long hot days, and the cares of city life."

The writer was referring to Saratoga Springs in New York state. There, the Eastern moneyed and middle class went to "take the waters." Not incidentally, they could also enjoy fashionable fine dining and elegant dancing. Nearby they could find top-notch horse racing and a fancy casino. After visiting Placerville and Quartzburg, the writer moved on to the Idaho City and the Warms Springs Resort. He called Warm Springs "the Saratoga of Idaho." Despite the hyperbole, he was partly right. In various forms, Warms Springs would be a popular resort for almost 140 years.

Early Placerville, Magazine Engraving

The springs, and other Boise Basin attractions, did not commonly draw a national or international clientele. Yet the newspapers offered a steady stream of articles about the region. Families and individuals traveled from all over southwest Idaho to hunt, fish, and camp there. Of course, most of the sojourners came from Boise City. The *Idaho Statesman* said (August 14, 1891), "A large number of our people are camped on More Creek and Crooked River, hunting and fishing."

The area also attracted those who wanted more civilized amenities. Under its "Society" heading the *Statesman* lauded (August 20, 1899) the wonderful "summer home of Mr. and Mrs. Woodburn, built in the pine and fir forest, one mile from Idaho City. … It certainly is one of the most artistic houses in the western mountains." Although rustic in décor, it also had unique touches: "One specially beautiful thing in the house is the mantle made of carefully selected pieces of quartz, some of them throwing streaks of gold. The mantle was designed and made by Mrs. Woodburn."

Top-Level Rustic Idaho Home, ca 1911

Born in Indiana, Harry Woodburn made his mark in business as a fairly young man. He and his wife Genevieve began visiting Boise, apparently about 1896. At that time, Harry owned a fence company in Minneapolis. Then, from the *Statesman* we learn (May 12, 1897), "Mrs. Harry Leonard Woodburn will recite an original poem at the Presbyterian church tonight."

The following year, Harry bought several important placer and quartz mining properties near Idaho City. He was just forty years old. Presumably he and Genevieve built their rustic home during this period. Woodburn sold the mining properties, and the home, in 1910. Within a year or so, they moved to the East Coast. Genevieve admitted, in letters to old friends, that she "was homesick for Idaho."

Before he left, Woodburn introduced the region to the quintessential machine of the new century. The *Idaho Statesman*, reported (October 27, 1904), "Harry L. Woodburn has purchased an auto that is adapted to mountain roads. It is the buckboard model and runs up steep grades with ease. Two Harry L.'s, Woodburn and Fisher, enjoyed the first ride in the new auto."

Of course, we met attorney and mining investor Harry L. Fisher back in Chapter Five. From the remaining description – a two-seater that cost $500 – the car was almost certainly an "Orient Buckboard," manufactured by the Waltham Manufacturing Company. They touted the Orient as the "best-selling Automobile in the world," able to "climb all grades up to 25 per cent."

1904 Orient Buckboard Automobile

Within two years, Boiseans owned "probably 15 or 20" automobiles and there was talk of forming an auto club. The *Statesman* said (May 28, 1906), "There have been a few new machines purchased this spring and the owners are enthusiastic over the sport afforded by the gasoline vehicle."

Not until 1913 did Idaho require owners to obtain a state license for their vehicles. Drivers themselves did not need a license until 1935.

As they moved into a new century, Boise residents, in particular, seemed to yearn more and more for a taste of "the great outdoors." Families went into the mountains to camp, and the husbands and sons found fishing spots to their liking. During a hot summer, the *Statesman* observed (July 26, 1905), "It is good to … imagine one's self

Fly Fisherman on a Western Stream

standing on the bank of some clear, cold mountain stream angling for trout. But it is better still to actually do the thing."

Recall that in July 1908 the Forest Service had restructured Idaho's forest reserves to create the Boise National Forest. A few years would pass before the Service began to directly encourage tourism in the Forest.

The *Statesman*, however, became an early booster for the region. On June 27, 1909, a writer said, "You can spend a day or two or even a week in casting the fly in some of the mountain streams near Boise for the speckled trout. It is sport for any king. Old Ike Walton never saw anything like it."

Earlier that same year, a stage line operator bought a Stanley Touring Car and

1909 Stanley Touring Car

tested the roads from Boise to Placerville and Idaho City. He hoped to run it as an "auto stage" in place of his horse-drawn coach. The route to Placerville was okay, but he had his doubts about making it to Idaho City. The *Statesman* report (March 15, 1909) said, "The run was without accident save the breaking of the pumping rod which necessitate the pumping of water into the boiler by hand."

Conditions improved and the line soon had regularly-scheduled auto stage runs into the Basin. Soon, ordinary motorists began making similar trips.

The newspaper tried to encourage this kind of motor tourism. They even went so far as to test likely routes. Thus, the *Statesman* described (August 6, 1916), two trips made by their reporters, "One of 132 miles which can be comfortably made in 12 hours any summer day with time for lunch and cooking coffee over a camp fire included, and the other of 90 miles, for which 8 hours is ample allowance of time."

Lunch Served Over a Quick Campfire

The following year, they began publishing illustrated stories about "outdoor Idaho" written by Otto M. Jones, who also took the accompanying photographs. The family moved to Idaho two years after Otto was born on a ranch near Dillon, Montana in 1886. He went to prep school out of state, and then lived for two years in Ashland, Oregon. That's when Jones began his career, writing about a wide range of outdoor sports.

Otto returned to Boise in 1909 and was married to a native Boisean two years later. Both Otto and his wife were outstanding shooters, placing high or winning in many city and regional matches. She also became his photographic assistant. Of course, they traveled all around Idaho, not just in the Boise River watershed. (Stanley Basin seemed to be a favorite spot.)

For several years, Otto's lively articles were a feature of the *Statesman's* Sunday edition. But he did more than just show pretty pictures. He

Writer-Photographer Otto Jones

told readers how to find the best spots for camping, hunting, and fishing. Jones also gave knowledgeable advice on camping equipment, cooking in the field, and much more. In his feature for July 21, 1918, he wrote, "If I were limited to only one cooking utensil in a kit for a mountain trip of any duration, I would select, without hesitation the Dutch oven for an all service utensil in camp."

Jones also sold articles to national magazines, such as *Field & Stream*. In 1917, that magazine featured a spread about a trip up the South Fork of the Boise River. The *Statesman* noted (March 25, 1917), "The article is illustrated by eight photographs taken by Mr. Jones, and should prove valuable as publicity for this section of the state."

Photographer Jones Captures Another Camera Enthusiast

Writing about the old mining camps, Jones said, "These fast disappearing camps fairly teem with sentiments and reveries for the traveler who halts long enough in his whirling pilgrimage to explore and conjecture as to the life of the ghost towns ... "

From 1919 into 1923, Jones served as the chief Warden of the Idaho Department of Fish and Game. During his tenure, the governor signed a new, comprehensive game law. Oddly enough, Idaho's first game law dated back to January 1864, when the Territorial legislature established guidance for the "preservation of wild animals." Of course, that law was mostly ignored, especially during the railroad building period. The law passed in 1919 made it illegal to kill game animals from February 1 through June 30. One could also not offer game meat for sale during that span.

By around 1920-1921, proselytizing for the streams, forests, and mountains had begun to pay off. Of course, the relatively low cost – gas, then cheap, and basic camping

Bird Hunters Posed with "Bag" by Jones. Shortest Woman May be Mrs. Jones

gear – help popularize such recreation. Thus, the U. S. Forest Service began taking a much more active role in promoting tourism in the national forests.

The *Statesman* said (June 19, 1921), "The forest service is boosting the recreational phase of forest work because it is quite generally recognized that the closer our people come to nature and wild life, in our mountain areas, the more they will appreciate all that forestry and forest protection means to our people."

As noted in the last chapter, Idaho and the nation were still trying to recover from the recession that followed World War I. In line with that, citizens started to pick up their travel and recreational activities. The *Statesman* observed (June 21, 1922) that "Fishing in various parts of the Basin is still good – if one knows where to look for the fish."

The newspaper also tried (June 30, 1922) to spur people to go: "To stimulate interest in finding attractive places in the foothills around Boise, *The Statesman* will give a $10 prize for the best account of a campers' trip to Bogus Basin, on July 2 and 3, and $1 for each kodak picture taken on the trip selected for publication in *The Statesman*."

Nice Trout!

For various reasons, several years passed before the Forest Service's attention to recreational activities had many tangible results. Up until about 1925, "improvements" mainly consisted of a dozen or so rudimentary campfire rings and some crude rest room facilities. Generally, hunters made their own way, but the lack of civilized amenities probably did discourage families and other casual vacationers.

Finally, the Service began to upgrade and add roads, and to establish new trails. Many of these efforts focused on popular geothermal springs, especially those close to roads suitable for cars. Of course, the commercial Warms Springs Resort continued to draw visitors.

Warm Springs Resort

These improvements and publicity campaigns did slowly increase tourist traffic into and through the Boise River watershed. Of course, at that time, logging and the mines were still important factors in the local economy. In fact, Idaho City actually showed a modest increase over the decade ... to about 190 in 1930. Also, the 1930 Census essentially reversed the occupational mix of 1920. That is, workers involved with logging or forestry substantially outnumbered those in mining, or any other occupation. Thus, while Basin stores welcomed the visitors and their business, residents made no particular effort to cater to them.

Those travelers surely meant business for the Luna House hotel, operated by Walter Galbreaith until his death in February 1925. Unfortunately, the Luna burned down in late 1927.

Luna House Hotel, ca 1920

Less than three years later, a new hotel, the Smith Hotel and Boarding House, opened in Idaho City. According to local lore, the structure contains timbers salvaged from a much older hotel in Placerville. The Smith accommodations generally served logging crews and forestry agents, but surely let the occasional room to a tourist.

During the Twenties and Thirties, hunting was probably the most popular "recreational" activity in the region. Of course, the intent changed after the stock market crash in 1929. Then, many people stayed alive with what the family hunter, or hunters could bring in.

Much of this is recorded only in oral histories (family stories) because many paid as little attention to games laws as they could get away with. In Idaho, at least, hunters often went out on horseback: Few people could afford to operate a car, even if they had one.

It is perhaps significant that Elizabeth Smith's *History of the Boise National Forest: 1905-1976,* says, "In 1930, a start was made in obtaining data for managing big game on the forest, and special concern was given to the winter range areas."

Hunting Party in the Idaho Forest

Wiegel's Soda Fountain and Store, ca 1930

One spot in Idaho City became a popular destination for hunters, campers, and other tourists. Recall that Melvin Wiegel had shut down his saloon operation when the state went dry in 1916. Even when the country repealed Prohibition in 1933, he did not, apparently, re-open a bar in his store. Hunters surely would have admired the décor, even if they couldn't buy an alcoholic drink there.

By the Thirties, administrators of the Boise National Forest had greatly expanded their programs. In 1929 and 1930, rangers studied factors that exacerbated erosion in the forest. Not only did erosion clog the streams and degrade fish habitat, it destabilized trees growing on the slopes. Dead and fallen timber greatly added to the fire danger in the forest. Other teams surveyed range areas commonly allotted for stock grazing. Managers needed solid data so they could work with stockmen to prevent over-grazing.

In 1933, the Forest Service created the Boise Basin Experimental Forest, containing tracts of carefully-controlled and monitored forest and range land. Here, foresters could refine techniques for timber management, soil and slope conservation, range management, and other matters.

Even more importantly, 1933 saw the creation of the national Civilian Conservation Corps (CCC). The CCC combined the Roosevelt Administration's desire to generate jobs, and a conservation ethic. Initially, the big labor unions opposed the Corps, seeing it as a training program that would produce more workers to compete for jobs with union members. Assured that young CCC enrollees would only engage in low-skill manual labor tasks, the unions went along. (The program eventually ignored the implied promise, and did train corpsmen for higher-skilled positions.)

As noted earlier, the rugged national forests in Idaho had seen very little campground development. That, plus the fact that the state contains *so much* Federal land, meant that Idaho received a disproportionate share of CCC contingents. They built camps all over the state, with many for the Boise National Forest. These included sites near Idaho City and Centerville.

CCC Camp Serving Boise National Forest

Those accepted into the CCC program were mostly "city boys" – young men aged 18 to 25 years from urban (mostly Eastern) environments. Their families were generally on some sort of relief program because no one could find a job. Of the young men's $30 monthly stipend, $22 (or $25, depending on various conditions) went home to the family. Still, the enrollees also received medical care, a clothing allotment, and board and room. They mostly had to built their own "room," of course.

After passing a rigorous physical examination, the young men headed to places none of them had ever heard of, much less seen. Those who arrived first discovered they had to live in big tents, in camps supervised by Army officers on detached duty or from the Reserves. A parallel command line of civilian supervisors and foremen did little to soften the impact. Under their leadership, the enrollees built their new "homes" – simple, open cabins with few amenities.

Thus, these young men experienced a major shock in moving from the cities to the wilds of Idaho. But most adjusted, and came to appreciate the fresh air, natural scenery, and serenity. They also gained a sense of confidence from their new physical fitness, and proven ability to complete strenuous tasks. A young man from New Jersey kept a journal about his experiences in Idaho. He wrote, "All in all this life seemed to agree with me. It was lonely, yet peaceful and soothing. It was a good change from the noisy city life."

The vast majority thrived, filled out, and found themselves with impossible appetites for the simple, but plentiful diet of the camps. After completing

CCC Workers Cutting Down a Big Tree

their cabins, they built roads, cleaned up streams damaged by slides, and planted seedlings to reforest barren hillsides. When necessary, they fought forest fires and rescued hapless tourists.

They also assembled the infrastructure to turn the national forests – including Boise National Forest – into America's playgrounds. Besides many roads and trails, they cleared campgrounds, built camping and picnic tables, and installed rest room facilities. A half century later, some of their tables were still in use. Even today, the campsites remain in place, and it is believed some of their original materials have been "recycled" into new structures.

Picnic Site Established by CCC Team

As the Depression waned, the CCC was glad to cut back its activities due to lack of recruits. As with so much else, World War II ended the need for the CCC as an emergency jobs strategy.

Still, largely as a result of their efforts, recreational visitation to the Boise National Forest increased by fifteen to twenty times compared to the 1920s. In fact, even with help from CCC crews, the Forest Service could not keep up. Soon, private interests began seeking permits to build facilities to meet the demand.

Although the War curtailed recreational activities, public and private developments laid the basis for a post-war surge. Indeed, Boise River gold country is still benefitting from the groundwork done at that time.

CHAPTER NINE

WORLD WAR, AND AFTERWARDS

The young man sipped his hot coffee while he flipped through the Sunday newspaper. The Sports Page should be a relief from stories about the draft, or fighting on the Russian Front. Hmm, Utah won the football game over Arizona down in Tucson. What's this deal about the Rose Bowl? He distantly noticed the end of some music show and wondered vaguely what came on next.

Then, a radio voice, authoritative but colored by suppressed tension, said "From the NBC news room in New York. President Roosevelt said, in a statement today, that the Japanese have attacked the Pearl Harbor … Hawaii, from the air. I'll repeat that. President Roosevelt says the Japanese have attacked Pearl Harbor in Hawaii from the air."

Befuddled patrons asked themselves if they'd heard aright. The station returned to its regular program, so someone twiddled the dial. Finally, they found a confirmation and discussion on another station. The common thread was, "We're at war!"

Same-Day Extra Announcing Attack

This scene is a composite of reactions all across the country. In many cities, newspapers rushed out same-day "Extra!" editions, or broadsides about the attack. Experts appeared on radio to pontificate about what it all meant. The following day, Roosevelt made his famous "live in infamy" speech, and Congress declared war on Japan little over a half hour later.

As has already been suggested, the U. S. entry into World War II profoundly impacted every aspect of Boise River gold country. Miner Louie Packer was among those affected. He had continued to work his claim right into 1941, although the family had relocated to a home in Boise. Louie also built a cabin on the claim. In a rough memoir, his daughter Irene said, "We continued spending our summers at our cabin in Idaho City for many years."

Packer Cabin on Spanish Fork

Family records show a regular succession of dates for assay reports from Idaho City or Boise through most of 1941. That included results on August 30, September 22, and October 20. Right after that, the memoir noted the Japanese attack on Pearl Harbor. Louie and his wife tried to sell the claim in 1945, apparently without success. Irene does not say what her father did during and after the war. Her mother went back to teaching school in Boise, in 1948.

In 1952, records show that they tried to buy some parts to repair a small prospector's mill. That same year, they had one last assay performed. The mine was inactive after that, and Louie died in 1960.

As noted in Chapter Seven, the War Production Board shut down gold mining in October 1942. After the war, many miners – like Louie Packer – chose not to go back to the mines. That seemed to have been particularly true for lode mining in the Boise watershed. Only a few small-scale operations, employing just a handful of men, are known to have started up again after the war.

Thus, mines in the Grimes Creek drainage (Pioneerville, Quartzburg, and so on), produced perhaps $100 thousand worth of gold in the fifteen to twenty years after 1940. (Moreover, some of that was probably from placer mining.)

The Ophir Group of lode mines, north of Rocky Bar, also re-opened as a very small operation. In the Fifties, the owner had just four to seven

Adit into Small Mine

men working under ground. Even that minimal activity ceased after about 1954-1955.

Companies at the quartz mines around Atlanta operated at a larger scale than the Ophir. Despite higher production with newer methods during the late Thirties, much rich ore remained. After the war, these mines and mills generated solid returns until about 1954. Then, they too shut down.

Large-scale placer mining did restart after the war in several parts of Boise River gold country. A dredge operated on the Middle Fork below Atlanta, and another obliterated the old Buena Vista camp near Idaho City. Still, most of this activity had ceased by 1958. Over its history, Boise River Gold Country produced about 3.3 million ounces of the precious metal. At current prices, that production would be worth over $5 billion.

Dredge Operating Near Idaho City

In the Seventies, a rise in the price of gold sparked some new interest in the old mines. Prospectors may have also searched for new possibilities, but the documentation is poor on that. In any case, not much came of this flurry of activity. Thus, in 1983, a historian of the Boise National Forest observed that only "recreational" miners were seeking placer gold in the area.

In total contrast to mining, Roosevelt's War Production Board *encouraged* logging and lumber production. In 1943, they initiated the Timber Production War Project. Sub-projects under that umbrella studied ways to use construction materials more efficiently, maximize lumber yields from cut logs, and enhance timber production from the forests.

Gold Mining the Old-Fashioned Way

They soon – it's not quite clear when – realized that none of that mattered if companies had no loggers. Some men had signed up, and many had been drafted, before the Board made logging a deferred occupation, vital to the war effort. Later (again, it's not entirely clear how much later), they extended that deferment to sawmill workers.

No one seems to have totaled up how many wooden structures the armed forces built during World War II. Private companies gearing up for war work added many more. Thus, the demand for lumber and structural timbers was enormous. For example, the Navy built seventeen huge wooden blimp hangars around the nation's coast. The one at Tillamook, Oregon – now an air museum – is still one of the largest freestanding wooden structures in the world.

Blimp Hangar Construction at Tillamook, 1943

The demand for timber products pushed the concern about forest fires even higher. Yet the loss of manpower to the military and expanded industrial demand left the Forest Service severely shorthanded. Once again, as they had during World War I, women began handling lookout duty and other jobs that would free up men for fire fighting.

The manpower shortage also forced the Service to move their experimental "smokejumper" project into operational practice. Better to drop a few men by parachute to fight a small fire than to wait hours or even days for a ground crew. They could at least hold the fire in check un-

Smokejumper Landing. Seems to be Older Style Chute

til ground reinforcements arrived. An early success in 1939 resulted in the formation of a small team the following year. Soon, the Service established a smokejumper base in Missoula, Montana.

Smokejumper with Packed Equipment Load

The crews found that "necessity was the mother of invention," as they modified equipment to meet their unusual needs. Weight was always a key issue. Today, potential candidates must pass a "pack out" test. That is, they must be able to cover a flat three-mile course in 90 minutes or less, carrying a 110 pound pack of gear.

Parachute professional Frank Derry made the most important early innovation. He devised a slotted canopy that opened more easily, was more stable in the drop, and was much more maneuverable.

The manpower problem eased somewhat when about sixty conscientious objectors were enrolled in the training program.

After the war, the recruitment crisis ended. Veterans filled a substantial number of the available positions. To improve response time, and relieve congestion at the Missoula site, the Service set up other bases, including one in Idaho City, in 1948. Jumpers would operate from there for over twenty years.

Of course, the smokejumpers were just the "shock troops" of the forces deployed to fight Idaho forest fires. Serious efforts to control fires in the Payette and Boise river watershed had begun in 1908. Guy F. Mains, a Forest Service supervisor, worked out a fire-protection agreement between the Service and representatives of state and private forest interests. In 1919, their understanding formally became the Southern Idaho Timber Protective Association (SITPA).

Southern Idaho Timber Protective Association Headquarters, McCall

For many years, the Forest Service deployed seasonal employees who maintained roads and trails, and otherwise kept busy, until they had to fight a fire. As time passed, the Service improved the quality and quantity of equipment available, and more and more fire watch towers went up.

Addition manpower would be hired for really large fires, almost literally "off the street." Although every regular Forest Service employee was expected to pitch in, not all of them had in-depth training in fire control and suppression. That changed for the Boise National Forest (BNF) in 1955, when the Service began training complete crews and assigning them to specific forests during fire season.

Timber production remained flat, or increased only slowly across the state for much of the decade after the war. Conversely, the allowed cut in the Boise National Forest increased dramatically from 1947 (21.5 million board feet) to 1952 (38 million). Then, after an extensive multi-year survey, the BNF experienced a "step function" in the allowed cut ... to 129.9 million board feet in 1956. Comparable increases were allowed on many Federal lands across the state.

The Boise-Payette Lumber Company, originally created in 1913, had remained the major producer in the BNF. In 1957, they merged with the Cascade Lumber Company of Yakima, Washington. The Boise Cascade Corporation would predominate in the Boise National Forest for the rest of the Twentieth Century.

To reduce the "collateral" impact of logging on the forest, companies and the Forest Service began to devise less invasive ways to get equipment in and logs out. These changes were particularly important in the BNF, where much of the timber grows on moderate to very steep slopes.

Log Truck, 1950s

In 1959, operators began using various elevated cable techniques in the Boise National Forest. Several configurations are possible, so we won't discuss the details here. However, cut logs hung from the high cable can be dragged, or hauled free and clear, through the forest for long distances (possibly up to a mile). The basic goal is to reduce the number of logging roads needed for big trucks.

Later, loggers began to use balloons or helicopters to lift logs directly out of the forest, often referred to as "aerial logging." Of course, neither technique can handle really big logs, and balloons can only be used under favorable weather conditions. (In

Helicopter With Log in Tow

fact, balloon logging is virtually unknown today.) The heavy-duty helicopters used to lift logs burn a lot of fuel, so operators try to minimize the haul distance as much as possible.

By 1970, forest products income accounted for over half the payroll in Boise County. Despite one minor setback in 1976, that revenue increased steadily to a peak of around $43 million in 1980. That income plummeted during the national recession in 1982-1983. Unfortunately, forest-related income never recovered, and, in fact, fell steadily after that.

Over this time span, an individual who would play an important role in more recent Boise Basin history was working as a smokejumper at the Idaho City base. Born in Caldwell, in 1932, Kenneth R. Smith worked briefly for a seed company after graduating from high school. He then joined the U. S. Air Force and conducted mountain survival training during the Korean War.

Kenn joined the smokejumpers in 1955 and eventually rose to a Foreman's position. His crew flew out of Boise after the Forest Service consolidated the Idaho City base with the one in the capital in 1970. In 1969, he was injured in a helicopter crash in the mountains between Arrowrock and Anderson Ranch reservoirs. Kenn lived in constant pain after that, and retired from the smokejumpers in 1972.

Idaho City Smokejumpers. Kenn Smith, Middle-Row Left, Semi-Kneeling

After his retirement, Kenn became involved with matters in Idaho City, including activities to preserve its history. At some point, he began, or had already begun, to collect old-time bottles. He would ultimately own an extensive and unique collection. Kenn played a significant role in the restoration of several historical Idaho City structures, including the Boise Basin Museum, the Miners' Exchange Block, and the Boise County Courthouse.

Recall that the original structure that became the Courthouse had been built in 1871 by an owner who took pains to make it fire resistant (Chapter Three). A general store at first, a later owner turned it into a hardware and trinket shop. From there, it became a hotel, which the county bought in 1909 to use as a courthouse. The renovation in the 1980s included an interior remodel to restore some of its pioneer character.

A later news article (*Moscow-Pullman Daily News*, October 22, 1990) said, "Ken Smith, resident for 36 years, is known as 'Mr. Idaho City' for his involvement in community affairs." Kenn also served as President of the Idaho City Historical Foundation.

The Idaho City Historical Foundation had been organized in 1958, with a mission to "preserve, protect, and interpret the history of the Boise Basin now and for future generations."

The Foundation sought to save some of the oldest pioneer buildings from collapse, or being demolished for their materials. One such structure had been built by James Pinney right after the devastating fire of 1867 (Chapter Two). Pinney sold the property in 1873, shortly after moving to Boise. After about 1882, one family owned the structure for over sixty years. The next owner sold it to the Idaho City government in 1953.

Boise Basin Museum, 1976

Concurrent with organizing the Foundation, historically-inclined residents also began converting the old Pinney structure into a museum. The Boise Basin Museum opened in June 1959. As noted earlier, they later performed some major renovation work on the structure. As artifacts became available, and funding permitted, the Museum organized new exhibits and upgraded existing attractions.

With more features to draw visitors, the flow of historical tourism slowly increased. However, the recreational "bread and butter" for Boise River country for over a decade

Forest Service Shelter/Campsite, Boise National forest

after World War II was still hunting and fishing. Then, in the late Fifties, the Forest Service made a concerted effort to upgrade National Forest campgrounds to make them more family-oriented. Highway, road, and trail improvements also helped encourage vacationing campers.

One challenge in the late Fifties and early Sixties was the appearance of motorized off-road vehicles – trail bikes, and all-terrain vehicles. Even before Honda introduced its "sporty" ATV in 1970, entrepreneurs devised a variety of small off-road transporters. Initially, hunters and people who worked roadless, or nearly roadless, country purchased most such vehicles. (Some even built their own.)

Increasing use of these powered vehicles began to seriously degrade the National Forest trails. Soon, the Service had to devise new regulations to control this traffic. Also, they redesigned some trails to handle the harder wear and tear. Other accommodations were required as purely recreational uses overtook "utilitarian" applications. These measures became critically important as manufacturers introduced ever-more-powerful ATVs.

Honda® ATV, 1970 Model

Snowmobiles, in some form, were invented as far back as the 1930s. Of course, they did not begin to look like what we are used to until the late Fifties. It is not at all clear when the first "modern" machines made it to Idaho, and to the Boise National Forest. Still, their popularity here grew more or less in parallel with ATV use during this period. Again, the Forest Service had to adapt. And, obviously, so did residents in the Boise Basin.

"Vintage" Snowmobile

As noted earlier, timber-related income rose steadily during the 1970s. That provided more leeway for projects to preserve historically-significant artifacts and structures. Probably not coincidentally, researchers for the Federal "Historical American Buildings Survey" visited Idaho City. They recorded facts and photographed a number of buildings.

One of those was the Miners' Exchange Block. Like most structures in Idaho City, the Block had a series of owners over the years. That included Melvin Wiegel, whom we met in Chapters Five and Eight. Wiegel's Place continued to be a popular destination for hunters and tourists until he sold it in July 1945. Several more owners

Miners' Exchange Block, 1976

followed. The holders as of 1973 poured considerable money into a major renovation and restoration, especially for the interior. That face-lift, partly using a selection of Wiegel's stuffed animals, transformed "a dilapidated old building into a thriving commercial enterprise."

Miners' Exchange Cafe, 1976

Tourism and other recreational activities got a boost in 1975, when funding agencies combined to upgrade the main road into the Basin – State Highway 21 – to "highway standards." Perhaps significantly, the population of Boise County, which had hovered around 1,600 to 1,700 since the War, jumped to over three thousand in 1980. Also, in 1981, income from retirement and investment returns – on the rise – equaled the rapidly falling income from forest products.

As timber jobs disappeared (never to return), residents who wanted to stay in the area sought others ways to make a living. Where that has led will be explored in our final chapter.

CHAPTER TEN

IDENTITY **TBD**

I must admit that my wife and I did not personally see Idaho City until 2004, over thirty years after moving to Idaho. We had heard about, read about, and seen pictures of the town, but had never been there. We drove in on State Highway 21, from Lowman over Mores Pass, on a summer trip.

That stretch of road is very steep in places, and snakes crazily along and over high ridges. No surprise that it closes periodically during the winter due to storms and avalanche danger. During the Boise Basin heyday, only an alternately dusty or muddy track, replete with stones, ran over the Pass. How daunting must that route have been for horse-drawn vehicles?

Western Stagecoach in Forested Mountains

**Idaho City, 1939. Wiegel's sign visible, middle left.
Mining and timber still fueled the town's economy.**

Idaho City "announced itself" with glimpses of old dredger piles off to the side of the road, with more as we drew closer. Finally, we turned up Main Street and checked out the town. On the weekday we visited, we saw few other tourists. Working people entered and left many buildings, most of which were obviously very old.

At that time, we did not know that much about Idaho City, other than that it had been a gold town and was the county seat of Boise County. We looked in at the Boise Basin Museum. That proved to be very interesting, with fascinating artifacts and great period photographs. Plus, the attendant mentioned some of the other historically significant landmarks we should know about. We had heard about the pioneer Roman Catholic church, so we made a point of visiting St. Joseph's.

Episcopal Church, Placerville

We next took the gravel road to Placerville. That too proved very interesting, although few of its early structures remain. However we did find the very well preserved Episcopal Church. We later learned that it had been built in 1894, with perhaps some later repairs or alterations.

Some years after that visit, Skip Myers sparked my interest in the history of the Basin and the Boise River gold country. As suggested in the earlier chapters, most of the remaining historically significant structures in Idaho City, Placerville, and other towns date back to the early days of the mining era. As mining waned, logging became the main economic driver. However, for various reasons, that too declined drastically. Recreation and tourism grew some, but they have not been a major source of income.

Boise Basin Mercantile, ca 1900. Built in 1865. Sold general merchandise until 2002. Has since been a gift shop, antique store, and more. Currently a saloon.

So the question must be asked: What now? As it happens, while I pursued the research for this book, the Boise County Commission was also seeking an answer to that question. Of course, roughly half of Boise River Gold Country lies in Elmore County. However, only about two hundred people live within that half, so we can focus more on Boise County. In any case, the dominant factor for the entire region is the presence of the Boise National Forest.

As mentioned in Chapter Six, Congress created the national forests to oversee their use in a sustainable manner. A 1976 revision re-emphasized a multiple-use approach. One provision required that "the public lands be managed in a manner which recognizes the Nation's need for domestic sources of minerals, food, timber, and fiber from the public lands ... "

However, the 1976 revision also included stipulations that the lands be managed "in a manner that will protect the quality of scientific, scenic, historical, ecological ... " and other values. That pushed the Forest Service into a difficult balancing act. Unfortunately, scientific forestry soon took a back seat to other agendas.

In 1987, the Forest Service commissioned an assessment of multiple-use issues for Region 4, which included the Boise National Forest. The report contains a telling statement: "It is, however, difficult to think of anything that has created more difficulties for Region 4 in particular and the Forest Service in general over the years than political conflict."

Critics – from all over the country, not just Idaho – charge that under such political pressures the Service has now largely abandoned multiple-use in favor of protection. The numbers for Idaho support that view. From 1948 through 1995, Federal forest lands in Idaho produced an average of about 670 million board feet of lumber annually. That was remarkably similar to the average production for privately-held timber lands.

For 1996 through 2006, production on private lands increased slightly. Yet over that period, output for Federal forests in Idaho dropped to about 115 million. For sixty years, privately-held forests have *sustained* an average cut of about 690 million board feet annually. With almost four times the acreage, Federal lands now allow only about one-sixth of that amount.

CCC crew clearing a service road in the Boise National Forest, ca 1938.

The *Boise County Comprehensive Plan*, issued by the Commission in 2010, acknowledged the drastic decline in Federal timber production. The negative impact is even more painful because the Boise National Forest takes up almost three-quarters of the county's area. The report says, "The trend seems to be warning Boise County to expect a dramatically decreasing level of revenue."

So, short of a major turn-around in Federal land-use policies, logging is a dead end for future job growth. Grazing and farming have never been an important source of

Chinese miners with sluice box, 1871. Chinese miners and merchants played a significant role in Boise River Gold Country for almost a half century.

Basin employment. So, in terms of producer-type jobs, we must revisit mining. After all, as the photo on the back cover shows, a lot of gold is still out there to be found.

Yet mining is, for all intents and purposes, now virtually a non-factor in the economy of the region. The *Comprehensive Plan* included employment data through 2007. Their table showed just 23 people engaged in mining, out of a total employment of over 2,300. The *Plan* briefly notes a "long history" of mining in the county. However, the planners do not mention any future possibilities.

The *Plan* Introduction contains this, slightly clarified, statement: "It is the intent of the County Commissioners to use the plan to guide future land use decisions that will promote a healthy living environment, [provide] an economic climate beneficial to the county, and reflect the character of the county desired by the residents."

Given the dismal history of ecological degradation by large-scale mining, that activity certainly can be incompatible with "a healthy living environment." Moreover, while the gold discoveries virtually "made" the Boise Basin, some residents do not want mining to be part of "the character of the county" today.

That feeling goes at least as far back as the Depression years. Recall from Chapters Seven and Nine that Louie Packer took up gold mining in the Basin, desperately hoping to survive the Great Depression. In her memoir, his daughter recalled, "Our Idaho City teacher made no bones about being against gold miners and, of course, pointed Marjorie and Irene out to the other kids as being gold miner's daughters."

(One can feel for the little girls held up to rebuke and ridicule in this way.)

Clearly, commercial gold mining is unlikely to provide many new jobs in the area. Still, opportunities do exist for small operations, if they are properly managed. At the end of 2011, prospectors had over two hundred active placer and lode claims in the Boise Basin. Maps also identified another hundred-plus abandoned claims. Beyond that, there are thousands of acres of unclaimed public land.

But in the end, the *Comprehensive Plan* leaves little choice but to pursue recreation and tourism as a source of future job growth. Yet decisions about recreational opportunities are largely out of their hands. Most of the best resources are inside the Boise National Forest, controlled by the Forest Service.

Guard Station, Boise National Forest

To draw tourists, Basin residents must exploit their colorful gold mining history. That need can lead to tension and discomfort at two levels.

Those with a deep-seated antipathy to mining want to apologize for the environmental damage done, not treat the miners as admirable, hard-working pioneers. While perhaps understandable, such an attitude is unreasonable; the present has no control over the past. Those men, and a few women, braved the privations and dangers of a primitive wilderness to build communities. We can surely celebrate their courage and determination.

But beyond that, long-time residents often resist the whole idea. Some of that tension surfaced in 1995. Historic preservationists were striving to acquire the house of Pon Yam, a prominent Chinese businessman who flourished during the height of Idaho City's gold boom. (They finally did, and it is currently being restored.)

The article in the *Deseret News* (November 24, 1995) about the Pon Yam effort mentioned the on-going battle: "The old-timers speak with some scorn of modern buildings in sight of Highway 21 dressed up to look like the real thing."

But the push for tourist-friendly venues worries even those who accept the need. They fear that the true, down-to-earth character of the town could be lost to shops full of kitsch and phony re-creations. That worry is not unfounded, yet the degradation is not inevitable.

Just last year, the City of Idaho City addressed some of the potential problems. In September, the City Council issued an *Idaho City Historic District Design Guide*. The *Guide* details the styles, materials, signage, landscaping, and other features that are appropriate to the city's traditional Historic District. The *Guide* categorically states that, "To achieve a sensitive recreation of the historic character, an awareness of the Historic District design features is essential. Whether for repairs, expansion, or new construction, the city's traditional building forms and materials must be respected."

City ordinances require that owners obtain a "Certificate of Appropriateness" *before* starting any project that would alter the exterior appearance of any structure in the historic district. The same goes for landscape changes, such as a fences, signs, lawn features, and so on.

By now, the source of this Chapter's title should be clear: The future identity of the Boise River gold country is still "to be determined." The county's *Comprehensive Plan* defined the stakes. Without local jobs for newcomers, and for young people coming of age, the Commissioners feared that "the area will end up as a 'bedroom community' to the Treasure Valley." While perhaps not a tragedy, that would be a shame.

Miners' Exchange Saloon, with pioneers out front. After 1890,
but before Milton Wiegel purchased the property.

IMAGE SOURCES

Abbreviations and Number References

AC : Author's collection.

CVM : Chippewa Valley Museum, Eau Claire Wisconsin.

HTF : Hiram Taylor French, *History of Idaho: A Narrative Account of Its Historical Progress, Its People and Its Principal Interests*, Lewis Publishing Co., Chicago (1914).

ICHF : Idaho City Historical Foundation.

IDC : Idaho Department of Commerce.

IL-ST : *An Illustrated History of the State of Idaho,* The Lewis Publishing Company, Chicago (1899).

ISHS : Idaho State Historical Society.

JHH : James H. Hawley, *History of Idaho: The Gem of the Mountains, The S. J. Clarke Publishing Company, Chicago (1920).*

LoC : Library of Congress.

NA : National Archives.

NOAA : National Oceanic and Atmospheric Administration.

NPS : National Park Service.

OHS : Oregon Historical Society.

SmI : Smithsonian Institution.

UI : University of Idaho Historical Archives.

USFS : U. S. Forest Service

USGS : U. S. Geological Survey.

YVHS : Yakima Valley Historical Society.

[a] Walter B. Stevens (ed.), *Centennial History of Missouri*, The S. J. Clark Publishing Company, Chicago (1921).

[b] Gabriel Franchére, Hoyt C. Franchére (ed. and translator), *Adventures at Astoria, 1810-1814,* University of Oklahoma Press (1967).

[c] McClurg Museum, Chautauqua County Historical Society, Westfield, New York.

[d] Alexander Ross, Kenneth A. Spaulding (ed.), *The Fur Hunters of the Far West,* University of Oklahoma Press, Norman (1956).

[e] Sandra Ransel, Charles Durand, *Crossroads: A History of the Elmore County Area,* Elmore County Historical Research Team, Mountain Home, Idaho (1985).

[f] Dan De Quille (William Wright), *History of the Big Bonanza*, American Publishing Company, Hartford, Connecticut (1876).

[g] W. A. Goulder, *Reminiscences : Incidents in the Life of a Pioneer in Oregon and Idaho*, T. Regan Publisher, Boise (1909).

[h] Lewis E. Aubury, *Gold Dredging in California,* Bulletin No. 36, California State Mining Bureau, San Francisco (May 1905).

[i] "The Hamilton Surface Planer," *Scientific American*, Munn & Company, New York (November 21, 1874).

[j] Alexander Toponce, *Reminiscences of Alexander Toponce*, University of Oklahoma Press, Norman (1971).

[k] Stewart Campbell, *Twenty-ninth Annual Report of the Mining Industry of Idaho for the Year 1927,* Idaho Bureau of Mines and Geology, Moscow, Idaho (1928).

[l] *History of Idaho Territory: Showing Its Resources and Advantages ...* , Wallace W. Elliot & Co., San Francisco, California (1884).

[m] "The Springs" Resort -- current web site.

[n] James D. "Smokey" Stover, Idaho City Smoke Jumper Foreman.

Image List (Page, Source. FC, BC = front and back covers)

FC	LoC		27	ICHF
i	USGS		28	IL-ST, ISHS 81-31-7
ii	AC		29	ISHS-76-119.2a , [e]
iii	LoC		30	ICHF, LoC
1	[a]		31	HTF, ISHS 1037-20
2	LoC, SmI		32	ISHS-66-23.2, HTF
3	LoC, AC		33	LoC
4	[b], [c]		34	LoC, HTF
5	[d], AC		35	AC, LoC
6	LoC		36	LoC, LoC
7	YVHC, LoC		37	AC
8	AC		38	LoC
9	ISHS-1254d		39	[g], LoC
10	OHS, JHH		40	ISHS-1246a , JHH
11	ICHF, LoC		41	ISHS-442c5
12	LoC, LoC		42	LoC, ISHS-81-68.8
13	LoC		43	HTF, ISHS-1333A
14	HTF, LoC		44	AC, IL-ST
15	LoC, ISHS-61-180.1 (G-1; A-R)		45	ISHS 61-15-2, ISHS-74
16	LoC, NA		46	ISHS-2094-11
17	NPS, [e]		47	HTF, ISHS-1037.22
18	ISHS 608-A, LoC		48	[e]
19	NA, [f]		49	ICHF, AC
20	LoC, HTF		50	ISHS-585h, HTF
21	JHH		51	LoC
22	Arn Hincelin Painting		52	LoC
23	ICHF, ICHF		53	NA, ICHF, ICHF
24	ISHS-78-193.1 , JHH		54	ISHS 62-20-28300b, HTF
25	JHH, LoC		55	LoC
26	ISHS-3308		56	ICHF

57 ICHF, ISHS 73-149-0	98 [e]
58 AC	99 LoC
59 ISHS 72-114-2/A	100 [k]
60 ISHS 73-59-2, ICHF	101 ISHS 62-50-2
61 [h], State of Idaho	102 [e], ISHS MS 2-1384
62 LoC, HTF	103 ISHS MS 2-1384
63 LoC	104 AC, LoC
64 AC, LoC	105 LoC
65 ISHS 60-120-1, ICHF	106 [l]
66 ISHS-62-86.8	107 ICHF, ICHF
67 AC, JHH	108 LoC
68 HTF, ISHS 65-134-1	109 AC
69 ISHS 73-129-9	110 LoC, AC
70 AC, HTF	111 LoC, JHH
71 [e]	112 LoC
72 [e], JHH	113 LoC, AC
73 AC	114 [m]
74 ICHF, LoC	115 ISHS 363b, NA
75 ICHF, LoC	116 LoC, USPS
76 ICHF, ICHF	117 USFS , ISHS MS683
77 UI	118 USFS
78 LoC, HTF	119 AC
79 LoC	120 ISHS MS 2-1384
80 ICHF, HTF	121 ISHS 63-160-3, LoC
81 NA	122 ISHS MS511-177-1a, NA
82 LoC, [i]	123 LoC, USFS
83 HTF, JHH	124 USFS
84 [e], AC	125 USFS, USFS
85 NOAA, ISHS 74-87.5	126 [n]
86 LoC, NA	127 LoC
87 LoC, LoC, LoC	128 USFS, AC
88 LoC	129 NA, LoC
89 AC, CVM	130 LoC
90 HTF, [j]	131 LoC
91 AC, LoC	132 ISHS MSS11-177-1e
92 AC, NA	133 LoC
93 AC	134 USFS
94 USFS, AC	135 LoC
95 LoC	136 USFS
96 AC	137 LoC
97 IDC	BC AC

BIBLIOGRAPHY

Books, Reports, Other Documents

Thomas G. Alexander, *The Rise of Multiple-Use Management in the Intermountain West: A History of Region 4 of the Forest Service*, Report FS-399, U. S. Forest Service, U. S. Government Printing Office, Washington, D. C. (May 1987).

William F. Bailey, *History of Eau Claire County Wisconsin,* C. F. Cooper & Co., Chicago (1914).

Hubert Howe Bancroft, Frances Fuller Victor, *History of Washington, Idaho, and Montana: 1845-1889,* The History Company, San Francisco (1890).

Merrill D. Beal and Merle W. Wells, *History of Idaho,* Lewis Historical Publishing Company, Inc. New York (1959).

Earl H. Bennett, *The Geology and Mineral Deposits of Part of the Western Half of the Hailey 1°×2° Quadrangle, Idaho,* U.S. Geological Survey Bulletin 2064-W, U. S. Department of the Interior, Washington, D. C. (2001)

John Bertram, Ellen Fenwick, *Idaho City Historic District Design Guide,* City of Idaho City, Idaho (September 2011).

"Post Office Block (Boise Basin Museum)," *Historic American Buildings Survey,* HABS ID-15, Department of the Interior, Washington, D. C. (1974).

Boise Basin Historical Summary, Boise County Government, Idaho City, Idaho (2010).

"Boise Basin Mercantile Company Block," *Historic American Buildings Survey,* HABS No. ID-13, Department of the Interior, Washington, D.C. (1980)

Marion Clawson, *The Bureau of Land Management,* Praeger Publishers, Santa Barbara, California (1971).

Comprehensive Plan Committee, *Boise County Comprehensive Plan, 2010 Update*, Boise County Board of Commissioners, Idaho City, Idaho (May 24, 2010).

Philip S. Cook, Jay O'Laughlin, *Idaho's Forest Products Business Sector: Contributions, Challenges, and Opportunities,* Policy Analysis Group – College of Natural Resources, University of Idaho, Moscow (August 2006).

Crofutt's Transcontinental Tourists' Guide, George A. Crofutt, Publisher, New York (1872).

Arif Dirlik (ed.), Malcolm Yeung (asst.), *Chinese on the American Frontier,* Rowman & Littlefield Publishers, Inc., New York (2003).

Elmore County, Idaho: 2004 Comprehensive Growth and Development Plan, Elmore County Commission, Mountain Home, Idaho (August 2004).

"Emmanuel Episcopal Church," *Idaho Episcopalian,* Vol. 2, Issue 2, Episcopal Diocese of Idaho, Boise (Winter 2011).

Encyclopædia Britannica from Encyclopædia Britannica 2007 Ultimate Reference Suite (2008).

Fire & Aviation Management, U. S. Forest Service (2011).

Philip L. Fradkin, *Stagecoach: Wells Fargo and the American West,* Simon & Schuster, New York (2002).

Gabriel Franchére, Hoyt C. Franchére (ed. and translator), *Adventures at Astoria, 1810-1814,* University of Oklahoma Press, Norman (1867).

Hiram Taylor French, *History of Idaho: A Narrative Account ...* , Lewis Publishing Co., Chicago and New York (1914).

"The Galbraith House," *Historic American Buildings Survey,* HABS No. ID-7, Department of the Interior, Washington, D.C. (1980).

W. A. Goulder, *Reminiscences : Incidents in the Life of a Pioneer in Oregon and Idaho,* T. Regan Publisher, Boise, Idaho (1909).

John Hailey, *History of Idaho*, Syms-York Company, Boise, Idaho (1910).

Arthur A. Hart, *Basin of Gold: Life in Boise Basin, 1862-1890,* Idaho City Historical Foundation, Idaho City (© 1986, Fourth printing 2002).

James H. Hawley, *Eleventh Biennial Report of the Board of Trustees of the State Historical Society of Idaho*, Boise (1928).

James H. Hawley, *History of Idaho: The Gem of the Mountains,* The S. J. Clarke Publishing Company, Chicago (1920).

"History of the District of Idaho," *U. S. Marshals Service,* United State Department of Justice.

Sherry Horton, *Moses Splawn, A Pioneer Portrait,* Idaho State Historical Society, Boise (1996, 2000).

Wilson Price Hunt, Hoyt C. Franchére (ed. and translator), *Overland Diary of Wilson Price Hunt,* translated from the original French *Nouvelles Annales des Voyages* (Paris, 1821), Ashland Oregon Book Society (1973).

An Illustrated History of the State of Idaho, The Lewis Publishing Company, Chicago (1899).

Washington Irving, *Astoria, or Anecdotes of an Enterprise Beyond the Rocky Mountains,* G. P. Putnam and Son, New York (1868).

A. H. Koschmann, M. H. Bergendahl, "Boise County, Idaho Gold Production," *Principal Gold-Producing Districts of the United States,* U.S. Geological Survey, Government Printing Office, Washington, D. C. (1968).

N. P. Langford, *Vigilante Days and Ways,* Montana State University (1957). Original publication in 1890.

Lawrence L. Loendorf, Nancy Medaris Stone, *Mountain Spirit: the Sheep Eater Indians of Yellowstone*, University of Utah, Salt Lake city (2006).

John F. MacLane, *A Sagebrush Lawyer,* Pandick Press, New York (1955).

"Masonic Temple, Idaho Lodge No. 1," *Historic American Buildings Survey*, HABS No. ID-8, Department of the Interior, Washington, D.C. (1980)

W. J. McConnell, *Early History of Idaho,* The Caxton Printers, Caldwell, Idaho (1913).

George A. McLeod, *History of Alturas and Blaine Counties, Idaho,* Hailey Times Publishing, Hailey, Idaho (1930).

"Miners' Exchange Block," *Historic American Buildings Survey,* HABS No. ID-14, Department of the Interior, Washington, D.C. (1974).

Victoria E. Mitchel*l, History of the Belshazzar and Mountain Chief Mines, Boise County, Idaho,* Staff Report 08-3, Idaho Geological Survey, University of Idaho, Moscow (July 2008).

Malcolm Rosholt, *Wisconsin Logging Book, 1839-1939*, Rosholt House, Rosholt, Wisconsin (June 1985).

Peter Skene Ogden, T. C. Elliott (ed.), "Peter Skene Ogden's Journal - Snake Expeditions," *Quarterly of the Oregon Historical Society* (1909-1910).

Edwin H. Peasley, *Twelfth Biennial Report of the Board of Trustees of the State Historical Society of Idaho*, Boise (1930).

Elias D. Pierce, as told to Lula Jones Larrick, J. Gary Williams and Ronald W. Stark (eds.), *The Pierce Chronicle,* Idaho Research Foundation, Inc., Moscow, Idaho (1975).

Sandra Ransel, Charles Durand*, Crossroads: A History of the Elmore County Area*, Elmore County Historical Research Team, Mountain Home, Idaho (1985).

Alexander Ross, T. C. Elliott (ed.), "Journal of Alexander Ross, Snake Country Expedition, 1824," *Quarterly of the Oregon Historical Society*, Vol. 14 (Dec. 1913).

Clyde P. Ross, *Mining History of South-Central Idaho,* Pamphlet 131, Idaho Bureau of Mines and Geology, Moscow, Idaho (July 1963).

Osborne Russell, Aubrey L. Haines (ed.), *Journal of a Trapper,* University of Nebraska Press, Lincoln (1965).

Sage Community Resources, *Payette River Scenic Byway Corridor Management Plan,* Idaho Department of Transportation, Boise (2001).

Elizabeth M. Smith, *History of the Boise National Forest: 1905-1976,* Idaho State Historical Society, Boise (1983).

Harold K. Steen, *The U. S. Forest Service: A History,* University of Washington Press, Seattle (1976).

United States Department of the Interior (eds.), *The Federal Land Policy and Management Act of 1976: As Amended,* U.S. Department of the Interior, Bureau of Land Management, Washington, D.C. (2001).

Irene V. Victory, *The Education of Gold Miner's Daughters: 1930-1943*, Idaho State Historical Society, Boise (1996).

Irene V. Victory, L. L. Packer, *Gold Mining in the Boise Basin, 1924 to 1958,* Idaho State Historical Society, Boise (1996).

War Production Board Limitation Order L-208, 7 Fed. Reg. 7992-7993 (Oct 8, 1942, with subsequent amendments).

Merle W. Wells, *Gold Camps & Silver Cities: Nineteenth Century Mining in Central and Southern Idaho*, 2nd Edition, Bulletin 22, Idaho Department of Lands, Bureau of Mines and Geology, Moscow, Idaho (1983).

Merle Wells, Arthur A. Hart, *Idaho: Gem of the Mountains,* Windsor Publications, Inc., Northridge, California (1985).

Gerald W. Williams, *The USDA Forest Service – The First Century*, Office of Communications, U. S. Department of Agriculture, Washington, D. C. (2005).

Oscar O. Winther, *The Great Northwest: a History,* Alfred A. Knopf, New York (1955).

Jim Witherell, *The Log Trains of Southern Idaho,* Sundance Publications, Ltd., Denver (1989).

World Almanac and Book of Facts, 2006, World Almanac Books, New York (2007).

Idaho State Historical Society Monographs

"The Boise City Assaying and Refining Works," Reference Series No. 3, Idaho State Historical Society (1962).

"Territorial Government In Idaho, 1863-1869," *Reference Series No. 48,* Idaho State Historical Society (1963).

"Boise-Idaho City Toll Road (1866-1902)," *Reference Series No. 78,* Idaho State Historical Society.

"Route of Alexander Ross, 1824," *Reference Series No. 86,* Idaho State Historical Society (July 1990).

"The Goodrich Trail," *Reference Series No. 93,* Idaho State Historical Society (April 1972).

"South Boise Wagon Road," *Reference Series No. 94,* Idaho State Historical Society (1964).

"Rice and Porter Toll Road (1868-1888)," *Reference Series No. 95,* Idaho State Historical Society (1983).

"Pacific Northwest Boundaries, 1848-1868," *Reference Series No. 104,* Idaho State Historical Society (1963).

"Alturas County," *Reference Series No. 112,* Idaho State Historical Society (May 1966).

"Census of 1863," *Reference Series No. 129,* Idaho State Historical Society.

"Census of 1864," *Reference Series No. 130,* Idaho State Historical Society.

"Stage Lines – Overland and Kelton," *Reference Series No. 146,* Idaho State Historical Society.

"Lumber in the Boise Region," *Reference Series No. 173,* Idaho State Historical Society.

"Site Report – Boise Basin," *Reference Series No. 198,* Idaho State Historical Society (1992).

"Idaho's First Year, 1863-1864," *Reference Series No. 226,* Idaho State Historical Society (February 1964).

"The Creation of the Territory of Idaho," *Reference Series No. 264,* Idaho State Historical Society (March 1969).

"Fort Boise (United States Army)," *Reference Series No. 356,* Idaho State Historical Society (1979).

"James Henry Hawley," *Reference Series No. 407,* Idaho State Historical Society.

"J. Marion More: Idaho Mining Pioneer (1830-1868)," *Reference Series No. 455,* Idaho State Historical Society (July 1994).

"South Boise Stage Lines," *Reference Series No. 465,* Idaho State Historical Society (1983).

"John Hailey: August 29, 1835-April 10, 1921," *Reference Series No. 543,* Idaho State Historical Society (1971).

"George Anislie (October 30, 1838-May 9, 1913)," *Reference Series No. 566,* Idaho State Historical Society (1981).

"William Henry Ridenbaugh, April 17, 1854-August 17, 1922," *Reference Series No. 594,* Idaho State Historical Society (1981).

"Albert H. Robie, 1830Õs - July 26, 1878," *Reference Series No. 596,* Idaho State Historical Society (1981).

"Alexander Rossi, March 10, 1828 - February 22, 1906," *Reference Series No. 597,* Idaho State Historical Society (1981).

"Location of Fort Boise and Boise City," *Reference Series No. 1119,* Idaho State Historical Society (June 1996).

Newspaper Articles

Many newspaper articles are quoted, and cited, in the book text. The following references provided information about key events, but were not directly quoted.

"Resources of the Basin," *Owyhee Avalanche*, Silver City, Idaho (November 3, 1875).

W. A. Goulder, "The Statesman On Its Travels – Centerville," *Idaho Statesman,* Boise (August 8, 1876).

W. A. Goulder, "The Statesman On Its Travels – Pioneerville," *Idaho Statesman,* Boise (August 15, 1876).

W. A. Goulder, "The Statesman On Its Travels – Idaho City," *Idaho Statesman*, Boise (August 5, 1876).

"Idaho City – Its Birth, Marvelous Growth, and Present Condition," *Idaho Statesman*, Boise (March 29, 1889).

"Mining News from the Basin," *Idaho Statesman*, Boise (November 24, 1889).

"Rocky Bar in Ashes," *Owyhee Avalanche*, Silver City, Idaho (September 10, 1892).

"Boise Basin Gold Mines," *Idaho Statesman*, Boise (October 12, 1893).

"One Dredge Completed," *Idaho Statesman*, Boise (August 14, 1898).

"Timber Trespass," *Idaho Statesman*, Boise (October 4-5, 1900).

"Boise Basin News," *Idaho Statesman*, Boise (December 19, 1901).

"Buys Valuable Mine Property," *Idaho Statesman*, Boise (June 3, 1903).

"Government Again Loses," *Idaho Statesman*, Boise (September 27, 1903).

"Sold to Wisconsin Firm," *Idaho Statesman*, Boise (April 7, 1904).

"Wealth of the Atlanta Mines," *Idaho Statesman*, Boise (October 11, 1904).

"Woodburn Placer Mines in Boise Basin Sold … ," *Idaho Statesman,* Boise (April 22, 1910).

"Extensive Placer Operation," *Idaho Statesman*, Boise (May 3, 1915).

R. N. Bell, "Placer Ground Still Yields Good Values," *Idaho Statesman*, Boise (May 19, 1915).

"Nature's Trysting Places Call Boise Folk to Come," *Idaho Statesman,* Boise (July 29, 1917).

Otto M. Jones, "Idaho is Banner State of West for Motor Touring," *Idaho Statesman,* Boise (April 21, 1918).

"Idaho Mining Conditions Promise Much for 1922," *Idaho Statesman*, Boise (July 21, 1922).

"Monster Dredge … ," *Idaho Statesman*, Boise (August 27, 1922).

"Abandoned Mine Yields $40 Ore," *Spokesman-Review*, Spokane, Washington (July 2, 1936).

"Gold Tide Gains in Boise Basin," *Spokesman-Review*, Spokane, Washington (March 24, 1936).